ELSEVIER OCEAN ENGINEERING BOOK SERIES

VOLUME 6

WAVE ENERGY CONVERSION

Elsevier Internet Homepage:

http://www.elsevier.com

Consult the Elsevier homepage for full catalogue information on all books, journals and electronic products and services.

Elsevier Titles of Related Interest

Published	Forthcoming titles
WATSON Practical Ship Design *ISBN: 008-042999-8*	BAI Marine Structural Design *ISBN: 008-043921-7*
YOUNG Wind Generated Ocean Waves *ISBN: 008-043317-0*	KOBYLINSKI & KASTNER / BELENKY & SEVASTIANOV Stability and Safety of Ships - Volume 1 / Volume 2 *ISBNs: 008-043001-5 / 008-044354-0*
BAI Pipelines and Risers *ISBN: 008-043712-5*	McCORMICK & KRAEMER Ocean Wave Energy Utilization *ISBN: 008-043932-2*
JENSEN Load and Global Response of Ships *ISBN: 008-043953-5*	MUNCHMEYER Boat and Ship Geometry *ISBN: 008-043998-5*
SAYIGH World Renewable Energy Congress VII *ISBN: 008-044079-7*	OCHI Hurricane Generated Seas *ISBN: 008-044312-5*
TUCKER & PITT Waves in Ocean Engineering *ISBN: 008-043566-1*	PILLAY & WANG Technology and Safety of Marine Systems *ISBN: 008-044148-3*

Related Journals

Free specimen copy gladly sent on request. Elsevier Ltd, The Boulevard, Langford Lane, Kidlington, Oxford, OX5 1GB, UK

Applied Ocean Research
Advances in Engineering Software
CAD
Coastal Engineering
Composite Structures
Computers and Structures
Construction and Building Materials
Engineering Failure Analysis
Engineering Fracture Mechanics
Engineering Structures

Finite Elements in Analysis and Design
International Journal of Solids and Structures
Journal of Constructional Steel Research
Marine Structures
NDT & E International
Renewable Energy
Renewable and Sustainable Energy Reviews
Ocean Engineering
Structural Safety
Thin-Walled Structures

To Contact the Publisher

Elsevier welcomes enquiries concerning publishing proposals: books, journal special issues, conference proceedings etc. All formats and media can be considered. Should you have a publishing proposal you wish to discuss, please contact, without obligation, the publisher responsible for Elsevier's civil and structural engineering publishing programme:

James Sullivan
Publishing Editor
Elsevier Ltd
The Boulevard, Langford Lane
Kidlington, Oxford
OX5 1GB, UK

Phone:	+44 1865 843178
Fax:	+44 1865 843920
E.mail:	j.sullivan@elsevier.com

General enquiries, including placing orders, should be directed to Elsevier's Regional Sales Offices – please access Elsevier homepage for full contact details (homepage details at the top of this page).

ELSEVIER OCEAN ENGINEERING BOOK SERIES

VOLUME 6

WAVE ENERGY CONVERSION

ENGINEERING COMMITTEE ON OCEANIC RESOURCES

Working Group on Wave Energy Conversion

(Chair: John Brooke, Vice-President, ECOR)

OCEAN ENGINEERING SERIES EDITORS

R. Bhattacharyya

US Naval Academy,
Annapolis, MD, USA

M.E. McCormick

US Naval Academy,
Annapolis, MD, USA

2003

ELSEVIER

Amsterdam - Boston - Heidelberg - London - New York - Oxford
Paris - San Diego - San Francisco - Singapore - Sydney - Tokyo

ELSEVIER SCIENCE Ltd
The Boulevard, Langford Lane
Kidlington, Oxford OX5 1GB, UK

First edition 2003

Library of Congress Cataloging in Publication Data
A catalog record from the Library of Congress has been applied for.

British Library Cataloguing in Publication Data
A catalogue record from the British Library has been applied for.

ISBN: 0 08 044212 9

⊗ The paper used in this publication meets the requirements of ANSI/NISO Z39.48-1992 (Permanence of Paper).
Printed in Hungary

SERIES PREFACE

In this day and age, humankind has come to the realization that the Earth's resources are limited. In the 19th and 20th Centuries, these resources have been exploited to such an extent that their availability to future generations is now in question. In an attempt to reverse this march towards self-destruction, we have turned our attention to the oceans, realizing that these bodies of water are both sources for potable water, food and minerals and are relied upon for World commerce. In order to help engineers more knowledgeably and constructively exploit the oceans, the **Elsevier Ocean Engineering Book Series** has been created.

The **Elsevier Ocean Engineering Book Series** gives experts in various areas of ocean technology the opportunity to relate to others their knowledge and expertise. In a continual process, we are assembling world-class technologists who have both the desire and the ability to write books. These individuals select the subjects for their books based on their educational backgrounds and professional experiences.

The series differs from other ocean engineering book series in that the books are directed more towards technology than science, with a few exceptions. Those exceptions we judge to have immediate applications to many of the ocean technology fields. Our goal is to cover the broad areas of naval architecture, coastal engineering, ocean engineering acoustics, marine systems engineering, applied oceanography, ocean energy conversion, design of offshore structures, reliability of ocean structures and systems and many others. The books are written so that readers entering the topic fields can acquire a working level of expertise from their readings.

We hope that the books in the series are well-received by the ocean engineering community.

Rameswar Bhattacharyya
Michael E. McCormick

Series Editors

v

Members of the Engineering Committee on Oceanic Resources (ECOR) Working Group on Wave Energy Conversion[1,2,3,4]

Chair: John Brooke, Vice-President ECOR, Dartmouth, Canada

Vice-Chair: Teresa Pontes, Instituto Nacional de Engenharia e Tecnologia Industrial, Lisbon, Portugal

Steering Committee :   Lars Bergdahl, Chalmers University of Technology, Göteborg, Sweden
                          Johannes Falnes, Norwegian Univ. of Science & Technology, Trondheim, Norway
                          Kenji Hotta, Nihon Univ., Chiba, Japan
                          Tom Thorpe, Energetech Australia Pty., Randwick, Australia

Members: Neil Bose, Memorial Univ. of Newfoundland, St. John's, Canada
                          George Hagerman, Virginia Tech Research Institute, Alexandria, USA
                          Hideo Kondo, Coastsphere Systems Institute Co. Ltd., Sapporo, Japan
                          Michael McCormick, U.S. Naval Academy, Annapolis, MD, USA
                          M. Ravindran, National Institute of Ocean Technology, Chennai, India
                          Zhi YU, Zhong Shan Univ., Guangzhou, China

Reporting Editor: Brian Nicholls, Bedford Institute of Oceanography, Dartmouth, Canada (*Retired*).

---

[1] The coordinates of the Working Group Members and the Reporting Editor are provided in Appendix 1.

[2] The Engineering Committee on Oceanic Resources (ECOR) is an international organization whose purpose is to provide an international focus and forum for ocean engineering activities, and to further international engineering activities pertaining to the management and exploitation of oceanic resources. In pursuit of this purpose, it undertakes a broad technical program in the field of ocean engineering through working groups or other kinds of subsidiary bodies, either alone or in conjunction with other appropriate organizations.

[3] The address of the ECOR Secretariat is: c/o The Royal Institution of Naval Architects, 10 Upper Belgrave Street, London, SW1X 8BQ, United Kingdom. Tel: +44 (0) 20 7201 2407. Fax: +44 (0) 20 7259 5912. E-mail: <rnicholls@rina.org.uk>

[4] For comments, updates and errata pertaining to this book go to <http://www.rina.org.uk> and select "ECOR".

ACKNOWLEDGMENTS

In addition to the members of the Working Group, the Engineering Committee on Oceanic Resources (ECOR) expresses its thanks to the following persons who provided information on wave energy conversion activities in their respective countries and/or details of specific systems, devices and other aspects:

- Max Carcas, Ocean Power Delivery Ltd., Edinburgh, UK
- Kyu-Bock Cho, Hanseo Univ., South Korea
- Alain Clément, Ecole Centrale de Nantes, France
- Steven Czitrom, National Univ. of Mexico, Mexico
- Don Dinn, Electrical Power Consultant, Dartmouth, Canada
- E. Friis-Madsen, Lowenmark F.R.I., Copenhagen, Denmark
- Fred Gardner, Teamwork Technology bv, The Netherlands
- Harry Hopf, U.S. Wave Energy, Longmeadow, Massachusetts, USA
- Seok Won Hong, Korean Research Inst. of Ships & Ocean Engng., Daejon, South Korea
- Terence Govender, EsKom Enterprises, South Africa.
- V. Jayashankar, National Institute of Ocean Technology, Chennai, India
- George Lemonis, Centre for Renewable Energy Sources, Pikermi-Attiki, Greece
- Jon Lien, Memorial University of Newfoundland, St. John's, Canada
- Mukhtasor, Sepuluh Nopember Institute of Technology, Surabaja, Indonesia
- Kim Nielsen, Danish Wave Power Aps / RAMBØLL, Copenhagen, Denmark
- Martin Renilson, Australian Maritime College, Launceston, Australia
- Charles Schafer, Bedford Institute of Oceanography, Dartmouth, Canada
- Hym-Jin Shim, Baek Jae Engineering, Seoul, South Korea
- Alf Simpson, South Pacific Applied Geoscience Commission, Suva, Fiji
- Arief Suroso, Sepuluh Nopember Institute of Technology, Surabaja, Indonesia
- Gareth Thomas, University College Cork, Cork, Ireland

The source(s) of material included in the figures and tables are identified. Every effort has been made to obtain permission to reprint such material. In the event of omission or error, the Reporting Editor should be notified.

PREFACE

"To leave no stone unturned"
-Euripides

Civilization over the centuries has not changed the human need of the physical requirements of heat and light for survival. However, population growth over the years has placed such a demand on the earth's primary available energy resources that there has been a constant search for additional sources to meet the increasing needs. Over and above basis needs, extraordinary demands for energy supply have increased many-fold due to the excessive and wasteful use of power by some countries, resulting in possible permanent damage to the earth's environment. The result is that present resources and methods of energy production, apart from the increased potential damage they may cause to the environment, may not be able to meet future world requirements. Therefore, alternative ways of supplying power must be developed and all possibilities explored. This need is more evident each day as the data concerning global warning is becoming more conclusive. Fortunately there is increasing evidence and practical proof that alternative energy sources can be gained by the use of the earth's natural energy sources that are both renewable and that have little impact on, and damage to, the natural environment. These energy sources are solar, wind and the ocean's potential energy, all of which have identifiable natural locations on the earth's surface, which enhance their performance. While individually each can provide only a percentage of the total requirements, together their output can supply considerable energy.

This book explores the potential of the ocean's energy from waves, an energy source known historically for its immense strength and destructive power. Yet this energy source, as this text will show, can be converted into useful work. Wave energy, together with other renewable energy sources, is expected to provide a small but significant proportion of future energy requirements without adding to pollution and global warming. The conversion of wave energy to power has various applications, and most naturally could find application in many coastal areas and islands, particularly those that rely on auxiliary diesel power stations, which are uneconomical and add to pollution. While some wave energy conversion methods are still at the early developmental stage, considerable progress has been made so that systems could now be installed on small islands and in other remote coastal communities. Energy costs, compared to conventional methods, are still high, but the history of technology has shown that these costs will come down considerably as in situ experience grows. Therefore it is anticipated that wave energy will ultimately supplement, or replace, a significant percentage of conventional energy sources, becoming competitive with present large-scale energy production methods without the associated problems of pollution.

The need for a review of potential methods of converting wave energy to useful power was brought to the attention of the Engineering Committee on Oceanic Resources (ECOR) by the late Dr. Nestor W. Lanfredi of Argentina via the Scientific Committee on Oceanic Research (SCOR). It is a mandate of ECOR to bring to the attention of the world in general, and government and industrial bodies in particular, developments such as wave energy: that is the purpose of this book.

This project would not have been possible without several contributing elements: the most important being the time spent by the contributors, primarily the members of the working group. It has absorbed many hours of work by these international experts, particularly those serving on the steering committee. Such effort is because these persons truly believe that this method of energy conversion must be used if the world is to be improved for coming generations. I also recognize the support of their employers who assisted them in both providing time and in meeting extra expenditures. Historically, I would also like to recognize the pioneering work of Yoshio Masuda of Japan, which provided early practical proof of this energy source.

Finally, but most important and particularly, I express thanks for the financial support provided by the Japan Marine Science and Technology Center (JAMSTEC). Without its aid, this book could not have been written.

On behalf of ECOR I thank them all for their help.

John Brooke, Working Group Chair, and Vice President, ECOR.

TABLE OF CONTENTS

Chapter 1

INTRODUCTION

The enormous energy potential of ocean waves has been recognized throughout history. However, it is only in recent times, following the oil crises of the 1970s when attention was focused on the possibility of extracting increased amounts of power from natural energy sources, that the exploitation of ocean waves in the production of electricity was explored in more detail. Experiments on wave energy conversion have indicated that several methods were feasible, and many areas of the world were shown to have the potential coastal wave energy that could be converted into useful power. The International Energy Agency (1994a) estimated that wave energy could eventually provide over 10% of the world's current electricity supply. Several wave energy prototypes have already been deployed worldwide (e.g., in Japan, Norway, India, China, UK, and Portugal {Azores}), and among demonstration and commercial designs are those being built in Australia, Ireland and the UK (Fig. 1.1). Support for these initiatives comes from governments, industry, investment companies and electrical utilities. Some small commercial companies are also developing demonstration wave energy devices.

With the exception of tidal waves, generated by the earth's rotation within the gravity fields of the sun and the moon, ocean waves appropriate to energy utilization are the product of surface winds. Of significance is that just below the ocean surface the density (or spatial concentration) of wave power is about five times larger than the corresponding wind power 20m above the sea surface, and 20 to 30 times the solar power.

Part of the solar energy received by our planet is converted into wind energy through the differential heating of the earth. In turn, part of that wind energy is transferred to the water surface, thereby forming waves. While the average solar energy depends on factors such as local climate and latitude, the amount of energy transferred to waves, and hence their resulting size, depends on the wind speed, the duration of the winds and on the distance over which it blows (know as "the fetch"). Strong winds must be present for an extended time to initiate fully developed wave generation, and these conditions must persist for regular periods throughout a year to create a useful source of energy. The most energetic waves on earth are generated between 30° and 60° latitude. However, there is also an attractive wave climate within ± 30° of the equator i.e. the "trade winds".

1

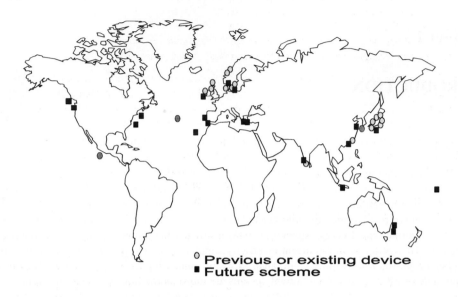

° Previous or existing device
■ Future scheme

Figure 1.1 Worldwide distribution of wave power schemes.
Source: T.W. Thorpe, Energetech, Australia. (Reprinted by permission.)

Although the early application of wave power has been restricted to coastal regions, as of 2002 deep offshore methods are being seriously considered. In coastal areas several factors influence the character of waves and hence their power. The type of the coastline and the near-shore bottom topography affect the energy available from waves at the coastline, i.e., a long shallow beach that extends gradually into the ocean results in energy loss as the waves dispel up the beach. Among other factors affecting the wave power available at the coast are wave breaking, coastal refraction (and diffraction) and sea bottom roughness.

The wave energy resource is usually expressed in terms of its power level, that is, the amount of energy available in each meter of a wave crest. The usual units of measurement are kilowatts per meter (kW/m) for a local area or megawatts per kilometer (MW/km) when referring to larger sub-regional areas. The global distribution of wave energy indicates that there are many countries that have a coastal wave climate favorable for exploitation of this resource. In these cases, the waves approaching each meter of coastline could satisfy the electricity demands of a number of households if a wave power system of reasonable efficiency was available (note that the average electricity demand of a household in a developed country is typically a few kW). The wave power resource at a location varies from day to day, from season to season, and from year to year, as do the wind and solar insolation. This variability in available power can be overcome by control systems that integrate conventional power systems for "back-up." Note, however, that waves are more persistent than winds, since there may be swells on the ocean when wind is absent.

Many of the systems noted in this book were largely developed during the oil crises of the 1970s. Research and experiments conducted to date place the technology at about the same status as flying was during the early 1920's; so its commercial viability is difficult to estimate at this time. It could be a period of ten to twenty years before substantial returns on investment are achieved but, like flying, early involvement is necessary to gain the knowledge and experience from research and development to enter the market place. What is certain is that eventually the power from renewable resources, such as waves, will play a major role in meeting world needs. An example of this is one of the earliest and most compact developments, by Yoshio Masuda of Japan, a wave-energy converter to generate power for a navigation buoy (see Chapter 8). Converters using his patent have been installed in over a thousand navigation buoys deployed world-wide.

This book details a variety of wave energy conversion systems. What follows is a simple catalogue of those systems and basic principles. Differences in system designs reflect the siting of units and the method of converting and transmitting the energy. However, there are variations within these divisions. Most designs convert the wave energy to electrical energy, but wave energy can also be used for the desalination of salt water by reverse osmosis, which is of vital importance for many societies and countries situated in arid climates. Basically there are three distinct categories: shoreline, such as built into a cliff face or rocky outcrop; near-shore; and offshore. The conversion of wave energy to usable power relies, in all cases, on the effect of the wave's force on some material or fluid, be that a movable mechanism or, for instance, air or hydraulic oil. Examples of this process include:

- *Water-to-air interface* – By allowing the wave action to enter a hollow structure, which has an air chamber above a movable body of water, the wave forces the air to move through an air turbine which in turn drives an electrical generator. This is known as an "oscillating water column" (OWC). It is the method that has been used most often for shoreline sites and at some near-shore applications, and was the method used in Masuda's light buoy.

- *Water action on movable bodies* – The wave action causes heaving, surging and/or pitching motion of a solid or flexible body (or linked bodies). This motion acts on a pump, forcing fluid through a hydraulic turbine or hydraulic motor.

- *Water storage method* – Waves are channeled along a tapered waterway on the coast, where sea water is lifted into a confined basin above the normal sea level. The water is then released back to the sea in a controlled sluiceway through a low-head turbine.

Presently all systems fall within these three fundamental approaches. Figures 1.2a, b and c show three possible versions. These and other designs are fully described elsewhere in this book.

3

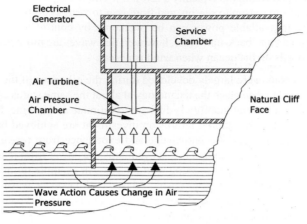

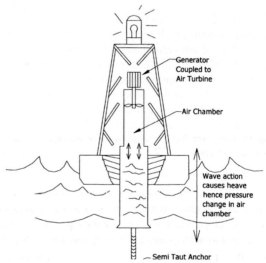

Figures 1.2(a) & (b) Simplified methods of wave energy conversion [(a)(top) – water to air interface; (b) (bottom) – water action on movable body]
Source: J Brooke, ECOR, Canada. (Reprinted by permission.)

4

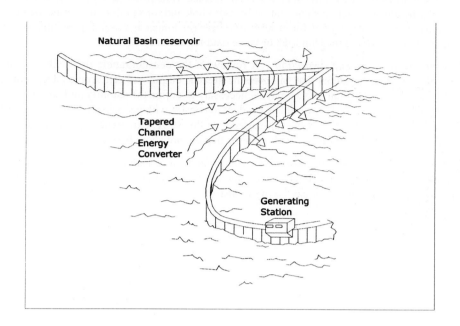

Natural Basin reservoir

Tapered
Channel
Energy
Converter

Generating
Station

Figure 1.2(c) Simplified methods of wave energy conversion (water storage)
Source: J. Brooke, ECOR, Canada. (Reprinted by permission.)

Wave power is less environmentally-degrading than most other forms of power generation, especially in relation to atmospheric emissions. Wave energy devices produce no gaseous, liquid or solid emissions and hence, in normal operation, wave energy is virtually a non-polluting source. However, the deployment and/or construction of wave power schemes can have some impact on the environment. Many of these potential impacts are site-specific and cannot be evaluated until a location for the wave energy scheme is chosen. If the site is carefully chosen (e.g. to avoid sensitive ecosystems), the environmental impacts are likely to be small.

With respect to the implementation of a wave energy project, the following are brief guidelines of the work that should be considered for design selection and site evaluation:

• *Oceanographic investigation* – A preliminary investigation of the proposed site(s) would involve reviewing existing oceanographic and wave atlases. In most cases, a hydrographic survey and site-specific positioning of current meters and surface wave recorders should be made only when the site for the plant has been selected. Data from these instruments would typically be transmitted to a shore station by radio link and /or satellite. It is important that these parameters be recorded for at least one year (or longer if possible), or linked to a longer-term record of either hindcast or wave data for a nearby site.

- *Design selection and site inspection* – A review of the oceanographic data will indicate possible conversion systems that will give the greatest output and efficiencies. Depending on the design, a geological survey may be necessary to define the stability and strengths of the underlining subsurface. It would be prudent to consult with geophysicists to understand the underlying structure below the sub-surface. At this stage, experienced ocean engineers should be consulted, particularly with systems that are not shore based.

- *Environmental and local considerations* – Environmental consultation must begin as early as possible. This includes reaching and gaining an understanding of the concerns of the local community, particularly the local fishing industry.

Chapter 2

OCEAN WAVES

2.1 Origin of Waves

Wave energy can be considered as a concentrated form of solar energy. Winds are generated by the differential heating of the earth and, as they pass over open bodies of water, they transfer some of their energy to form waves. Energy is stored in waves as both potential energy (in the mass of water displaced from the mean sea level) and kinetic energy (in the motion of the water particles). The amount of energy transferred, and hence the size of the resulting waves, depends on the wind speed, the length of time for which the wind blows and the distance over which it blows (the "fetch"). Power is concentrated at each stage in the transformation process, so that the original average solar power levels of typically ~ 100 W/m^2 can be transformed into waves with power levels of typically 10 to 50 kW per meter of wave crest length.

Waves lying within or close to the areas where they are generated appear as a complex, irregular "wind sea". These waves will continue to travel in the direction of their formation even after the wind is no longer acting on them. In deep water, waves lose energy only slowly, so they can travel out of the storm areas with minimal loss of energy, becoming progressively regular, smooth waves or "swell". These can persist at great distances (up to ten thousand kilometers or more) from the point of origin. Therefore, coasts with exposure to the prevailing wind direction and long fetches tend to have the most energetic wave climates, for instance the west coast of the Americas, Europe and Australia/New Zealand (see Fig. 2.7). The global wave power resource in deep water (i.e. ≥ 100 m water depth) is estimated to be $\sim 10^{12}$-10^{13} W, or 1-10 TW, (Panicker 1976).

2.2 Description of Waves

Sea waves can be simply described as a sinusoidal wave with the characteristics shown in Figure 2.1. A detailed description of waves is presented in Appendix 2 to this document, where a mathematical approach to wave statistics is outlined under two main topics: (a) hydrodynamics of sea waves; and (b) wave energy.

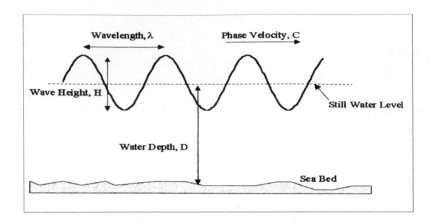

Figure 2.1 Description of a simple wave
Source: T.W. Thorpe, Energetech, Australia. (Reprinted by permission.)

2.3 Modification of Simple Waves

As waves approach the shore through waters of decreasing depth, in shallow waters (i.e. when the water depth is approximately half the wavelength or less) the sea bed begins to modify the waves in various ways:

- *Shoaling.* As the water depth decreases, an initial limited decrease in wave height is followed by progressive increases, first slowly and slightly, and then more rapidly. (Fig. 2.2a).
- *Wave breaking.* Steep waves break, thereby losing both height and energy (Fig. 2.2a). This is generally the mechanism responsible for the largest dissipation of the energy contained in ocean waves. Breaking can also have a positive effect in wave energy utilization in the near-shore (water depths less than 20-30 m), and especially at the shoreline, because the largest storm waves break before reaching a wave power generating plant, thus avoiding extreme loads on the structure due to waves breaking on it. Note also that in the near-shore zone the short waves of the spectrum start to be reduced by sea-bed interaction.
- *Bottom friction.* In shallow waters, the water disturbance caused by surface wave motion extends down to the seabed. In this case, friction between the water particles and the seabed results in energy loss that increases with the width of the continental shelf and sea bottom roughness (Fig. 2.2b).

- *Refraction.* Refraction, diffraction and reflection of ocean waves are similar to the corresponding optical phenomena. They are dependant on the detailed variation of the seabed topography, and promote the redistribution of wave energy density. As waves propagate into shallow waters, the wave fronts are bent so that they become more parallel to the depth contours and shoreline (Figs. 2.3 & 2.4). As a consequence, energy is concentrated in convex bathymetric formations such as submarine ridges and at headlands, leading in some cases to "hot spots" where remarkable concentration of energy occurs. Energy is dispersed where the sea bottom pattern is concave, e.g., in bay areas. Even in the absence of energy focusing or de-focusing, the change of direction due to refraction is of great importance to shallow-water wave energy devices whose capture efficiency is orientation-dependent.
- *Diffraction.* Among the effects due to this phenomenon is the bending of waves around and behind barriers. It promotes the transfer of energy from high- to low-energy concentration areas, thus generally being a negative factor for wave energy extraction. However, diffraction effects at indented coasts may cause energy concentration at the shoreline.

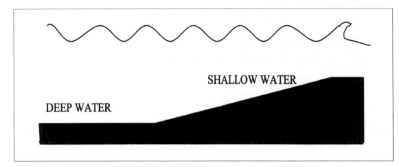

Figure 2.2a
Water transformation in shallow water: shoaling and br eaking
Source: Laboratorio Nacional de Engenharia e Tecnologia, Portugal 1993. (Reprinted by permission.)

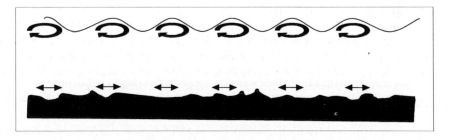

Figure 2.2b Wave transformation in shallow water: bottom friction
Source: Laboratorio Nacional de Engenharia e Tecnologia, Portugal 1993. (Reprinted by permission.)

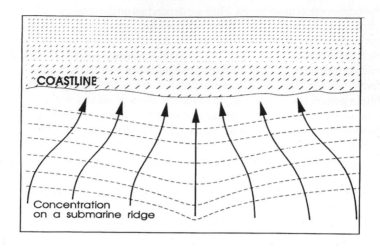

Figure 2.3 Wave transformation in shallow waters: energy focusing by refraction
Source: Laboratorio Nacional de Engenharia e Tecnologia, Portugal 1993 (Reprinted by permission.)

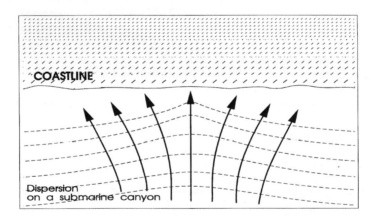

Figure 2.4 Wave transformation in shallow waters: energy de-focusing by refraction
Source: Laboratorio Nacional de Engenharia e Tecnologia, Portugal 1993. (Reprinted by permission.)

In the open ocean, wave conditions are approximately constant over distances of a few hundred kilometers in large ocean basins, such as the Atlantic and Pacific Oceans, and over distances of a few tens of kilometers in smaller basins, such as the Mediterranean Sea. However, in the near-shore, and especially at the shoreline, the wave energy resource can vary significantly over distances of 1 km, or much less, due to the occurrence of shallow-water phenomena and diffraction by the coastline, as well as due to the effect of shelter by neighboring islands. Various numerical models exist that compute the modifications caused by the occurrence of the above phenomena. The model types and their underlying assumptions are described in Southgate (1993). Pontes et al. (1993) surveyed the existing models and their applicability to wave energy resource assessment. Although recent increases in computer power have enabled significant increases in the accuracy of such models, and also the possibility to simulate complex phenomena such as breaking, no dramatic changes in shallow-water models have occurred since the studies referred to in the above references were undertaken.

2.4 Real Sea Characteristics

The above outline refers to the behaviour of simple, monochromatic waves. Real seas contain waves that are random in height, period and direction (Fig. 2.5). Within a short length of time, the characteristics of real seas remain constant, thereby comprising a sea state. In order to describe such sea states and to determine their characteristics relevant to wave energy devices, statistical parameters derived from the wave spectrum are used (see Appendix 2). The most usual wave height and period parameters are:

- Significant wave height H_s is defined as the average height of the highest one-third waves. This approximates to the height that a shipboard observer will report from visual inspection of the sea state, since such an observer tends to overlook the smaller, less conspicuous waves.
- Energy period T_e is a mean wave period with respect to the spectral distribution of wave energy transport (wave power level). Peak period T_p is defined as the period corresponding to the peak in the variance density spectrum of sea surface elevation. It thus represents the harmonic frequency component having the greatest amount of energy at a place passed by a random wave train. Some wave energy devices can be "tuned" to this frequency in a manner analogous to the tuning of a radio circuit in an electromagnetic wave field.

In deep water, the power in each sea state P is given by:
$$P = 0.5 H_s^2 T_e \quad \text{kW/m}$$
with H_s expressed in meters and T_e in seconds.

The annual variation in sea states can be represented by a scatter diagram (Fig. 2.5), which indicates how often a sea state with a particular combination of characteristic wave height and period occurs annually. Therefore the annual average wave power level (P_{ave}) can be determined from a scatter diagram as:
$$P_{ave} = \Sigma P_i W_i / \Sigma W_i$$
where sea states with power levels P_i occur W_i times per year. Based on the velocity of the wind that generated the waves, several models describing the wave energy density distribution in terms of frequency have been developed. One such model of sea states (Pierson & Moskowitz 1965) was often used in the United Kingdom Wave Energy Programme for programming wave makers for

11

model tests in wave tanks. The JONSWAP spectral model, named after the Joint North Sea Wave Project (Hasselmann *et al.* 1973), has become the most widely-used spectral model.

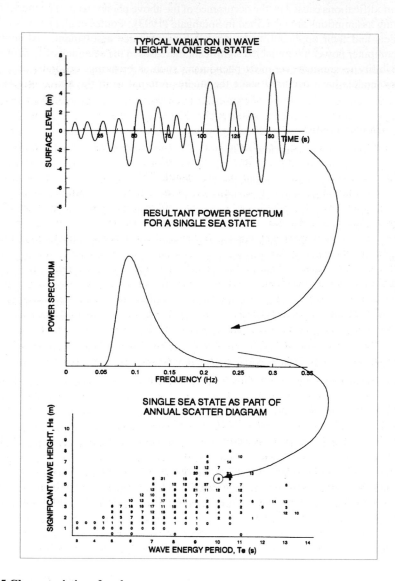

Figure 2.5 Characteristics of real seas
Source: T.W. Thorpe, Energetech, Australia. (Reprinted by permission.)

12

Variability

Although a statistically stationary sea state (i.e. waves neither building nor decaying) can be described by its significant wave height and dominant wave period, individual wave height and period vary more or less randomly from one wave to the next. Such variability is more random in seas generated by strong local winds and less random in swell that arrives from far distant storms.

Wave power levels change at all time scales, from wave to wave (order of seconds), over periods of hours to days (either in response to local wind conditions or the arrival of swell from distant storms), from season to season and year to year. Waves often occur in successive groups of alternately high and low waves, a phenomenon familiar to any surfer. The output of a wave energy device will tend to follow the envelope profile of such wave groups. Incorporating minutes-long storage of mechanical or fluid power (e.g., turbine flywheels or hydraulic accumulators) smooths these pulses, which is important from the viewpoint of the integration of the power plant into the electrical grid, and enables the use of lower capacity electrical equipment. This saving can offset the cost of such storage components, thereby improving overall system economics. The random seas built up by strong winds acting on a large fetch during a long period of time contain frequency components that can range from 0.04 to 0.25 Hz (period of 25 to 4 s). To a first approximation, the energy contained in each component travels independently, at a speed that is directly proportional to its period. These different components start out together upon leaving the area of active wind generation, but the long-period waves travel faster and soon outdistance their shorter-period counterparts. The energy of the fully developed storm sea is thus dispersed along a corridor emanating from the generation area. One consequence of wave dispersion is that a wave power plant located in the path of swell from such a storm will experience a rather abrupt increase in dominant wave period as the first, longest waves arrive, followed by a gradual decrease as the slower, shorter-period waves arrive. A second consequence of wave dispersion is that it lowers and broadens a storm's energy "pulse", enabling sustained recovery of this energy by a wave power plant located in distant waters.

Developing long-term statistics associated with the variability of wave-power levels is a key step in plant design, since it underlies the choice of geometry, the structural design and the selection of equipment rating. This involves a trade-off between under-utilization of installed capacity and excessive shedding of absorbed power (Fig. 2.6). In this figure, the top part of (b) represents typical output in mid-winter. The lower part represents typical output in mid-summer. In high latitudes, a plant designed for optimal year-round performance will shed much of its wave energy input during the winter, yet be significantly underutilized and perform at reduced efficiency during the summer. Although the annual average wave power input would be less in the tropics than in high latitudes, sea states are more consistent from day-to-day, possibly resulting in improved plant economics. Since the wave-power level is, for the most of the time, less than its average value, a design goal for the development of wave-energy converters will be that maximum annual energy production is obtained with lower power levels than indicated by the efficiency curve in this figure.

13

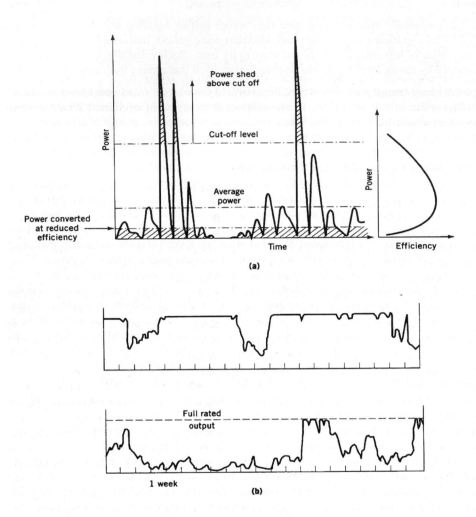

Figure 2.6 Day-to-day variation of: wave power input (a), and power plant output (b), in response to changing sea states
Source: Dawson 1979. (Reprinted by permission.)

14

Year-to-year variability in wave-power levels must be considered, particularly when developing resource data for economic projections. For example, annual average wave-power levels off northern California can vary up to 50% from one year to the next. Also, only one or two years of measured wave data may be available for a proposed plant site. In these cases, some attempt should be made to determine how well these data represent the long-term wave climate to which a plant would be exposed during its 20- to 30-year service life. This can be done by inspection of statistics obtained from long-term time series of wind-wave models results, of visual observations, or even of long-term wind data statistics, which are in general significantly correlated with the wave statistics, especially in wave generation areas.

2.5 Wave Climatology and Resource Distribution

The geographic distribution and temporal variability of wave energy resources are governed by the major wind systems that generate ocean waves: extra-tropical storms and trade winds. In some areas, notably India, local monsoons can also influence the wave climate. Extra-tropical cyclones are born when prevailing westerly winds off continental land masses pick up heat and moisture from western ocean boundary currents such as the Gulf Stream in the north Atlantic Ocean and the Kuroshio Current in the north Pacific Ocean. These low-pressure systems typically develop sustained wind speeds up to 25 m/s, blowing over a 1000 km fetch for two to four days before the storm makes landfall. Extra-tropical storms are most frequent and intense during the winter, when monthly average wave-power levels typically are three to eight times greater than summer monthly averages as (see the European Wave Energy Atlas, WERATLAS) (Pontes et al. 1996a). Where utility peak demand is dominated by winter heating and lighting loads (northern Europe, for example), wave energy has a good seasonal load match. Where peak demand is driven by summer air conditioning (California, for example), the seasonal load match is poor.

In the northern hemisphere, extra-tropical cyclones follow northeasterly tracks, continually building the waves in the storm's southern sector, which are traveling in the same direction as the storm. On the other hand, waves generated in the northern sector of a northern-hemisphere cyclone travel opposite to the direction of storm advance and have much less exposure to the storm's wind energy. Consequently, swell traveling "backwards" from such a storm has much less power than swell leaving the storm's southern sector. As a result, wave resources along the western part of an ocean basin are generally poorer than along the eastern part. In the North Atlantic Ocean, annual average wave-power levels along the edge of North America's eastern continental shelf ranges from 10 to 20 kW/m. By comparison, shelf-edge wave-power levels off the European western coastline increase from about 40 kW/m off Portugal up to 75 kW/m off Ireland and Scotland, decreasing to 30 kW/m off the northern part of the Norwegian coast. Figure 2.7 illustrates the general global distribution of coastal wave-power levels, while Figure 2.8 illustrates the spatial variation of annual mean power-levels, and the directional distribution, over the most energetic area of the northeastern Atlantic. A similar pattern occurs in the north Pacific Ocean. On the western side of the basin, off Taiwan and Japan, wave power averages 5 to 15 kW/m. Figure 2.9 maps power levels off the coasts of Japan. On the opposite side of the Pacific, off the coast of northern California, annual wave power-levels range from 25 to 35 kW/m.

In the southern hemisphere, extratropical low-pressure systems develop in the open ocean expanse that surrounds Antarctica. These storms travel from west to east, uninterrupted by land, and generate high-energy swells that make landfall along the southwest coasts of South America, Africa, Australia, and New Zealand. This swell also contributes significantly to the wave energy resources of Indonesia and island nations throughout the south Pacific Ocean. As in the northern hemisphere, annual power levels of incident waves decrease with decreasing latitude, averaging more than 100 kW/m just south of New Zealand, and dropping to 30-40 kW/m in deep water west of Auckland. Even further north, in the south Pacific region framed by Vanuatu, Cook Islands, Tonga and Tuvalu, wave-power levels at exposed island locations are estimated to be 15-20 kW/m. Figure 2.10 shows the distribution of annual wave-power levels close to the South Pacific islands of Fiji, Funafuti, Rarotonga, Tongatapu, Vanuatu and Western Samoa.

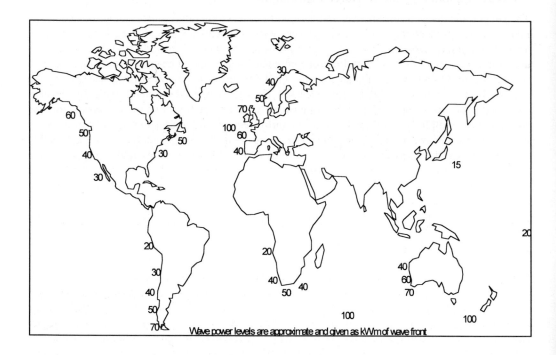

Figure 2.7 Approximate global distribution of time-average wave power
Source: T.W. Thorpe, Energetech, Australia. (Reprinted by permission.)

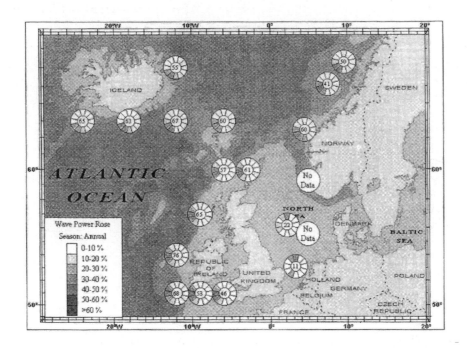

Figure 2.8 Annual wave power roses and mean gross power levels (kW/m) for the most energetic area of the northeastern Atlantic
Source: Pontes et al. 1996a. (Reprinted by permission)

Off tropical coasts, where trade winds are dominant, the annual offshore wave power level is in the range of 10 to 20 kW/m. Although trade winds never approach the intensity of extra-tropical storm winds, they are much more persistent and the seasonal variation is smaller than farther north or south. Windward island coasts in the tropics are affected not only by trade-wind waves, but also by swell from extra-tropical storms at higher latitudes. For example, the average wave-power level at exposed deep water locations in Hawaii is estimated to be 15 kW/m, divided about equally between northeast trade-wind waves and northwest Pacific swell (SEASUN 1992). Within a 15°(latitude)-wide belt centered around the Equator, tropical cyclones are absent, and the wave power level is mainly due to swells originating from storm regions farther north or south. From a technological point of view, there are good reasons for deploying wave energy converters at exposed sites within this belt, not because the wave power level is high, but because of the likelihood of avoiding problems associated with extreme waves.

17

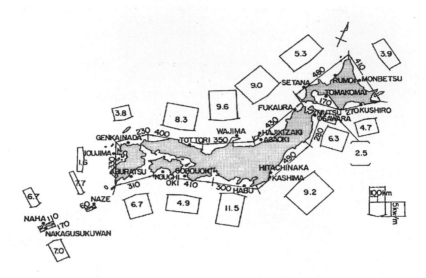

Figure 2.9 Annual wave power level (kW/m) off the coasts of Japan obtained from wave measurements at non-sheltered sites in water depths of 20 m or more. (The width of each arrow is proportional to the length of coastline to which the power level applies; the arrow length is proportional to the power level.). Source: Takahashi 1988.

2.6 Data

Two basic methodologies exist for the estimation of long-term series of wave data. The first is measurement and observation, while the second is based on building time series with numerical wind-wave models. A wide variety of in situ and remote sensing measuring methods are available that produce accurate wave data. Visual observations made aboard travelling ships are the earliest type of wave data for the world's oceans. Wind-wave models are computer programs that numerically generate and propagate wave energy based on input wind data. For open-ocean resource assessment in large ocean basins, such as the North Atlantic and Pacific Oceans, the accuracy is quite good. These models are implemented at most meteorological centers around the world. The collection of the available wave data sets is not an easy task because the data are archived at several institutions that have different procedures for dissemination.

18

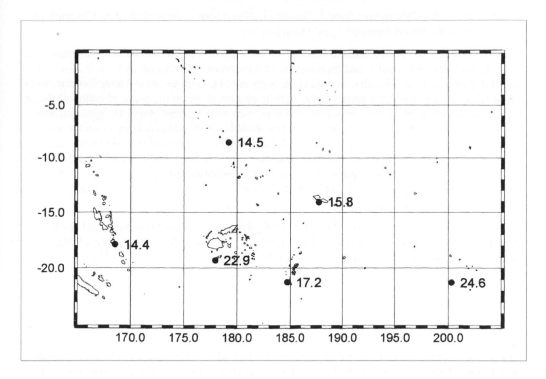

Figure 2.10 Average wave power levels (kW/m) measured by the wave-rider buoys deployed close to the South Pacific islands of Fiji, Funafuti, Rarotonga, Tongatapu, Vanuatu and Western Samoa within the SOPAC Wave Measurement Program

Source: South Pacific Applied Geoscience Commission (SOPAC), Fiji, and OCEANOR, Norway. (Reprinted by permission.)

Observations

The visual observations are made at synoptic times (0000, 0600, 1200, and 1800 Greenwich Mean Time) and include wave observations reported as the visually estimated height and period of both windsea and swell (World Meteorological Organization 1976). They have been archived by meteorological organizations from the mid 1850s onwards. Their storage has been committed to Germany (South Atlantic Ocean - including North Atlantic 0-20 degrees latitude), USA (western Pacific and western North Atlantic), United Kingdom (eastern North Atlantic), Netherlands (Mediterranean Sea, southern Indian Ocean), India (northern Indian Ocean), Hong Kong (South China Sea), Japan (eastern Pacific), Russia (seas adjacent to its northern coastline, southern hemisphere south of 50 degrees South). The data outlined above are published as Climatological Summaries of the World Meteorological Organization.

A serious problem with wave statistics based on visual wave observations is fair weather bias caused by ships avoiding bad weather and high waves. This problem is increasing as a result of increased use of ship routing to avoid high seas. The visually observed height, period and direction correspond respectively to the significant wave height, significant period and to the mean direction of the sea state. Calibration formulae have been proposed to convert the visually observed heights and periods into corresponding significant wave heights and periods. Several authors have assessed the accuracy of visual data. A review of such studies by Athanassoulis & Skarsoulis (1992b) revealed that visually obtained wave directions are most reliable, wave heights are considered satisfactory, but wave periods are less reliable. Visual wave observations are thus considered to be more appropriate for a comparative study of wave (and wind) climate for different areas or locations, and as a supplement to measurements.

Several global and regional atlases based on visual observations have been published, such as the global atlas by Hogben, Dacunha & Ollivier (1986). Athanassoulis & Skarsoulis (1992a) have developed a wind and wave atlas for the eastern Mediterranean.

Measurements

There is a wide choice of in situ and remote wave measuring systems available. The selection of a system depends on access, water depth, wave conditions at the measurement site, and the required data details, in particular whether directionality is required. Figure 2.11 schematically illustrates several measuring systems. Some devices provide directional information on their own or when coupled with other devices. In situ measuring devices can store the information or transmit it by cable, telemetry or satellite to a land-based processing station. Figure 2.12 shows an example of a system that allows local data storage as well as data transmission. Several attempts have been made at the difficult task of comparing device accuracy. One of the most significant of these was the WADIC project in the early 1980s (Allender et al.1989), which showed that several systems provide accurate results for present purposes. A summary of its conclusions is presented in Appendix 3. Within the Wave Crest Sensor Intercomparison Study (WACSIS), an intercomparison of wave measurements by various instruments also showed good agreement (Forristal et al. 2002).

In general terms, many wave measurements have been made for engineering purposes in specific studies, often with a limited duration of one year or less. These have usually been for coastal locations (e.g. for harbour construction, shore protection), and hence of only local value; and less often for the open sea (e.g. for wave climate purposes or at offshore oil platforms). For wave energy resource assessment, the most interesting sets of measurements (and the least common) are those taken over a long period, i.e., with a duration of several years, at non-sheltered offshore locations. Although such data are not abundant, they exist for most of the USA (including Hawaii), Canada and Europe (see Pontes et al.1993). Recently, more offshore measurements have been made, especially in the Mediterranean Sea.

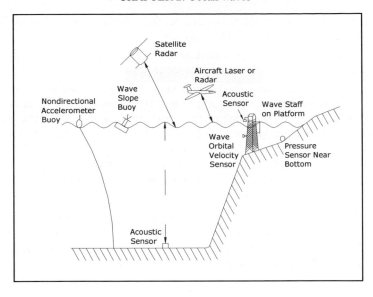

Figure 2.11 Schematic diagram of several wave measurement systems
Source: T. Pontes, INETI-Department of Renewable Energies, Lisbon, Portugal (adapted from Earle &
Bishop 1984). (Reprinted by permission.)

In-situ measurements in the open sea utilize mostly wave-recording buoys (scalar or directional).
Their results include time series of sea surface elevation (and tilt) from which wave height and
period (and direction) parameters can be derived by spectral or direct analysis of the time series.
Wave measurements close to the coast can be made by a number of devices besides buoys, including
submerged pressure and acoustic probes, wave staffs, current meters, orbital velocity sensors, as well
as acoustic probes suspended above the water surface. With the exception of directional buoys, if
these devices are used individually they provide nondirectional information, which is often
appropriate since refraction tends to reduce the wave directional spreading at the shoreline. When
grouped in arrays or coupled to another suitable instrument, for instance an electromagnetic current
meter, they can provide directional information. Other directional systems have been developed
based on the processing of video records of an array of floating elements. Direct measurements of
the oscillations of the water column inside the chamber of oscillating-water-column wave power
plants have been made using several device types. Acoustic probes inverted over the water surface
and bottom-mounted pressure probes are being used to monitor the LIMPET installation on the
Scottish island of Islay and the Pico Island plant in the Azores, Portugal. However, it should be noted
that in rough sea states such echo probes do not operate satisfactorily due to the spray that fills the
chamber. In the Japanese OWC wave power plant at the breakwater of Sakata harbour, wave staffs
were successfully employed.

21

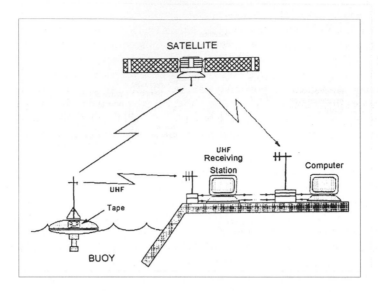

Figure 2.12 Sketch of an in situ wave measuring system based on a buoy with local data storage, as well as transmission to a shore-based processing station
Source: Laboratorio Nacional de Engenharia e Tecnologia Industrial, Portugal 1993. (Reprinted by permission.)

Remote sensing techniques provide spatial information about the sea surface rather than simply information at one point as given by the devices described above. There are several ways that wave conditions can be remotely measured without in situ wave sensors:

- Aerial photography, the simplest means to remotely sense waves, is useful to identify the occurrence of wave refraction and diffraction near coasts; such studies being useful for selecting shoreline and near-shore sites for power plant construction.

- Shore-based radar can be used for measurement of the waves approaching a wave energy power plant. Presently these radars are able to resolve direction with good resolution and have logistical advantages over in situ sensors.

- The most powerful remote sensing methods, however, are those utilizing satellites. There are two main types: the satellite altimeter and the Synthetic Aperture Radar (SAR). The first provides only the significant wave height; whereas SAR provides the directional wave spectrum for long period waves (above about 8-9 seconds and wavelength longer than about 100 m), but it does not detect wind sea.

The satellite altimeter is a vertically-pointing pulsed radar, developed for measuring the range to the Earth's surface. The satellite altimeter is capable of measuring significant wave height and surface wind velocity in most weather conditions. Models to compute wave period from altimeter data have been proposed (e.g Davies et al.1998) with further verifications being carried out, e.g., by Moreira et al. (2002). SAR has not been useful for wave energy resource assessment due to the requirement for high computing effort and the impossibility of detecting short waves. The high computing effort is needed to resolve directional ambiguity (e.g., if waves are oriented with propagation direction north-south it is not possible to distinguish between waves travelling towards and from the north). This requires starting the analysis with a first conjecture at the directional spectrum, usually using wave model results.

Although the feasibility of accurately measuring wave height from space was demonstrated by NASA's GEOS-3 satellite in 1975 and the SEASAT in 1978, it was with the launching of the US Navy's GEOSAT satellite in 1985 that a series of radar altimeters have been providing almost continuous (with a gap between 1989 and 1991) wind and wave measurements over the world's oceans. As of 2002, three satellite altimeters were operating globally. These are: the US Navy's GEOSAT Follow-On launched in 1998 (http://gfo.bmpcoe.org/Gfo); the US/French JASON (http://aviso.jason.oceanobs.com/), which followed the Topex/Poseidon launched in 1992; and the European Space Agency's ENVISAT (http://envisat.esa.int), which followed the ERS-1 mission started in 1991 and the ERS-2 launched in 1991. These satellites follow exact repeat orbits, i.e., the satellite returns to the same ground track after an "exact repeat period", which is 10 days for JASON, 17 days for GEOSAT FO and 35 days for ENVISAT. A global wind and wave atlas has been developed based on the data measured by GEOSAT (Young & Holland 1996).

Numerical wind-wave models

A wind-wave model numerically integrates the so-called energy balance equation, which expresses the budget of wave energy over a given area (usually represented by a grid point) and time interval (refer to Appendix 2) . Wind-wave models, which have as input the wind fields over the sea area of interest, provide estimates of directional spectra over a grid covering that area. These models have progressed through three "generations" in recent years, corresponding to increasing sophistication in the way that their equations represent numerically the physics of events. The last breakthrough occurred in 1988 with the publication of the first so-called third generation wave model WAM (WAMDI Group 1988). The novel characteristic of this model, since repeated in other models, was that no simplifying parameterization was used in the solution of the equations (which had been considered the case in the original formulations). A detailed description of wind-wave modelling, namely of the WAM model, can be found in Komen *et al.* (1994). Numerical meteorological models produce the input for wind fields, their quality being the key factor in determining the accuracy of wave model results. Waves are highly sensitive to small variations of the forcing wind field, thereby magnifying any small error. Accuracy also varies with the dimensions of the considered basin and with the complexity of the bordering topography.

Wind-wave models can be run in mainly three modes. When incorporating predicted wind fields as input, they produce wave forecasts. The forecasting period extends up to 10 days in the open ocean, but is smaller in enclosed basins because the accuracy of the input wind is inferior. Generally, the wave model is run afterwards, using analyzed wind, i.e., wind fields obtained from the forecast fields after the assimilation of measured (buoy and satellite altimeter) data. Major meteorological centers archive these wave analyses, which are the best possible estimates of past wave conditions. Hindcasts are the result of special runs of wave models for a specific time period. The so-called re-analyses are wave hindcasts computed by the most updated wave models, using driving winds obtained from the most recent meteorological models.

The WAM model is implemented in the routine operation of most major meteorological centres, in particular at the European Centre for Medium- Range Weather Forecasts (ECMWF), Reading, UK, where, in addition to the provision of daily wave forecasts, the wave analyses are archived. The development of the European Wave Energy Atlas (Pontes *et al.* 1996b; Pontes 1998) was based on the results of this model over most of the area covered. A verification over the northeastern Atlantic results for the period 1986-1994 against buoy and satellite altimeter data showed that there was good agreement between estimates and measurements for all parameters. As shown in Pontes *et al.* (1996b), on average the model slightly underestimated all parameters, but a higher underestimation occurred for some strong storms. A negative bias, smaller than 0.20 m for the significant wave height H_s (which corresponds to less than 10% of its mean value), was found; also the period parameters were underestimated by less than 5%. The error in the wave power level P is higher than for the individual height and period parameters because P is proportional to $H_s^2 T_e$. A negative bias of 10-20% of the average value was found for P. Subsequently, a significant increase in the accuracy of the ECMWF WAM model results has occurred due to improvements in the model itself, as well as in the accuracy of the input wind data.

Complementarity of data types

The relationship between collection methods and the types of data provided is indicated in Table 2.1, below. Satellites sample rapidly in time (1Hz.) along the ground track. An in-situ measuring instrument samples rapidly in time at a fixed point (typically 1Hz.) to provide measurements at, typically, 3 hr intervals. The global numerical wind-wave model estimates wave conditions at thousands of (grid) locations world-wide, typically every 6 hours. The main drawback of the satellite altimeter is the rather poor temporal resolution (the best being 5 days at the cross-over points of the Topex/Poseidon mission, but at the expense of larger, i.e., more coarse grid, spatial resolution). It turns out that the measurements and the global wave model complement one another very well, and are best used together rather than independently. This is shown in the validation work for the development of the European Wave Energy Atlas (Pontes et al. 1996b) where altimeter and buoy data were used to validate the numerical wind-wave model results.

24

DATA TYPE				
COLLECTION METHOD	Sea Surface Time History	Wave Spectrum	Sea State Parameters	Wave Statistics
Visual Observations			Primary data	Data reduction
Numerical Hindcast Models		Primary data	Data reduction	Data reduction
Remote Sensing Measurements		Primary data	Data reduction	Data reduction
In-situ Measurements	Primary data	Data reduction	Data reduction	Data reduction

Table 2.1 Types of data produced by various collection methods
Source: Hagerman 1992. (Reprinted by permission.)

2.7 Wave Energy Resource Assessment and Development

According to Hagerman (1992), wave energy resource assessment and development should proceed in three phases. The first two phases would probably be undertaken by a state government or utility concerned with a particular region. The third phase would be undertaken by a commercial developer at a specific project site. Phase I would make use of existing wave data (visual observations, hindcasts, measurements and visual observations, as available) to provide an initial assessment of the regional magnitude of the wave energy resource and its commercial potential. Depending on the quality and quantity of existing data, and the budget available for analysis, Phase I should also result in some indication of the geographic distribution of the resource, and its variability from season to season and from year to year. If Phase I results are sufficiently encouraging, then Phase II would involve selection of ten to twenty candidate sites for a 2- to 3-year numerical hindcast. Visual observation statistics would be used to select a hindcast period that is a good representation of long-term wave conditions. Selection of the candidate sites would consider the geographic distribution of the resource (as determined in Phase I), the type of wave energy technology being considered for development (e.g., land-based, near shore breakwater, or offshore), coastal topography, social and environmental concerns, and utility constraints, such as the degree of grid reinforcement required for transmitting wave power to load centres. In Phase II, the wave model would be calibrated with existing measured wave data. Thus in addition to the candidate sites for wave power development, existing wave measurement sites would also be included in the model's numerical grid. Although Phase II would not involve any new measurements, estimates of total wave power would be considerably improved over Phase I, and resulting cost-of-energy projections would be much more reliable. Economic assessments based on Phase II data would be adequate for commercial

25

developers to decide whether to build a wave power plant at one of the candidate sites. If a decision is to be made to build, then Phase III would involve the deployment of an in-situ wave measuring system at the project site and collection of measured data for a period of at least one year. The wave model would be further validated such that its output closely matches measurements made at the project site. A 20- or 30-year hindcast would then be performed, appropriate to the design life of the plant. Operational and extreme wave statistics (representing plant survival conditions) would be prepared from this long-term hindcast.

Chapter 3

WAVE ENERGY CONVERSION SYSTEMS

There are six major renewable energy resources in the oceans: waves, tides, thermal differences, ocean currents, salinity gradients and the biomass. The exploitation of ocean waves in the production of electricity and potable water is the primary interest of this book, and the different methods devised for capturing the energy of ocean waves are outlined in this chapter.

Examples of practical wave-energy devices that have been in use for decades are fog horns, bilge pumps and navigation buoys, the latter needing a battery to store energy for periods when wave activity is too small. Based on work by the Japanese wave-energy pioneer Yoshio Masuda, more than one thousand wave-powered navigation buoys have been produced and marketed worldwide since 1965; some have been in operation for more than 20 years. Air turbines are used in these devices for the secondary energy conversion, with the the power level being of the order of some hundred watts or less.

The basic wave energy conversion process can be stated in very general terms as follows: the force (or torque) produced in a system by an incident wave causes relative motion between an absorber and a reaction point, which acts directly on, or drives a working fluid through, a generator prime mover. The periodic nature of ocean waves dictates that this relative motion will be oscillatory and have a frequency range of 3 to 30 cycles per minute, much less than the hundreds of revolutions per minute required for economic, conventional electric-power generation. A variety of working fluids and prime movers are employed to convert these slow-acting, reversing wave forces into the high-speed, unidirectional rotation of a generator shaft.

3.1 Wave Energy Conversion Technology

Primary conversion of wave energy is attained by an oscillating system (either a floating body, an oscillating solid member or oscillating water within a structure). The system is able to store some kinetic and/or potential energy. A secondary conversion may be required to obtain some useful form of energy. In 19th century proposals, the energy associated with the oscillating motion was transmitted to pumps or other suitable energy converting machinery by mechanical means (such as racks and pinions, ratchet wheels, ropes and levers). In contrast, devices for control and power take-

27

off in modern proposals include controllable valves, hydraulic rams and various hydraulic and pneumatic components, as well as electronic hardware and software. Typically, secondary energy conversion is obtained by means of a turbine, which delivers energy through a rotating shaft. In addition, electric generators are included if the absorbed wave energy is to be converted into electricity (tertiary energy conversion).

The many different types of possible wave energy converters (WECs) may be classified in various ways, for instance, according to their horizontal size and orientation. If the size is very small compared to the typical wavelength, the WEC is called a *point absorber* (Budal & Falnes 1975). On the contrary, if the extension is comparable to or larger than the typical wavelength, the WEC is called a line absorber, but the terms "terminator" and "attenuator" are more frequently used. A WEC is called a terminator or attenuator if it is aligned along or normal to the prevailing direction of wave crests, respectively.

Another possibility is to classify WECs according to their different locations. They may be located on the shore, partly above and partly below the mean water level, or they may be completely submerged and placed on the sea bed, some meters below the mean water level. Otherwise a WEC may be moored in a floating position, or in a somewhat submerged position, either near-shore or offshore. A hybrid system is also conceivable, where one or several near-shore floating WEC units pump seawater (or another fluid in a closed loop) to an elevated water reservoir on land, from which a hydraulic motor (turbine) with an electric generator is run. If the land is low behind the shore, a pressure tank may be used as the reservoir.

Hagerman (1995a) identifies twelve distinct process variations, as shown in Figure 3.1. The classification system used in this book, which is a modified and updated version of Hagerman's system, is shown in Table 3.1. The main features that distinguish one process from another are mode of oscillatory motion for energy absorption, type of absorber, and type of reaction point. Energy can be absorbed from heave (vertical motion), surge (horizontal motion in the direction of wave travel), pitch (angular motion about an axis parallel to the wave crests), yaw (angular motion about a vertical axis) or some combination of these modes. Absorbers can be fabricated of rigid or flexible material, or can be the free surface of the water itself. Reaction points can be inertial masses (suspended plates, buoyant spines, or other absorbers), sea-floor anchors (deadweight or pile), or fixed, surface-piercing masses (concrete or land).

The phenomenon of primary conversion of wave energy may be described as follows. The wave force acts on a movable absorbing member, which reacts against a fixed point (land or sea-bed based structure), or against another movable, but force-resisting structure. Heave forces may, for instance, be reacted against a submerged horizontal plate. Wave forces may also be reacted against a long "spine" (this is common for several identical absorber units, and the length of the unit should be more than one wavelength). The wave force results in oscillatory motion of the absorbing member, and the product of wave force and corresponding motion represents absorbed (primary converted) wave energy.

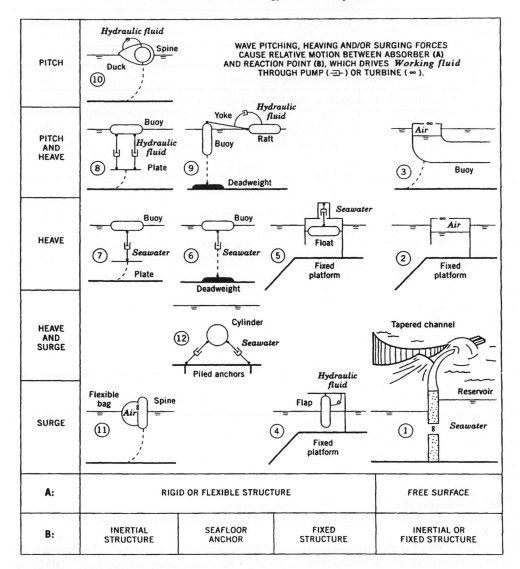

Figure 3.1 Classification of wave energy conversion processes based on mode of energy absorption (pitch, heave, surge or combined modes), type of absorber (A), and type of reaction point (B)
Source: Hagerman, G. 1995a. (Reprinted by permission.)

Primary Location of Device	Wave Energy Conversion Process		Hagerman #
	#	Title	(a)
Onshore	1.1	Fixed oscillating water column	2
	1.2	Reservoir filled by wave surge	1
	1.3	Pivoting flaps	4
Near-shore to Offshore	2.1	Freely floating oscillating water column	3
	2.2	Moored floating oscillating water column	3
	2.3	Bottom-mounted oscillating water column	2
	2.4	Reservoir filled by direct wave action	1
	2.5	Flexible pressure device	11
	2.6	Submerged buoyant absorber with sea-floor reaction point	12
	2.7	Heaving float in bottom-mounted or moored floating caisson	5
	2.8	Floating articulated cylinder with mutual force reaction	
Offshore	3.1	Freely heaving float with sea-floor reaction point	6
	3.2	Freely heaving float with mutual force reaction	7
	3.3	Contouring float with mutual force reaction	8
	3.4	Contouring float with sea-floor reaction point	9
	3.5	Pitching float with mutual force reaction	10
	3.6	Flexible bag with spine reaction point	11
	3.7	Submerged pulsating-volume body with sea-floor reaction point	

Table 3.1, Classification of wave energy conversion processes
(Key: (a) Cross reference to classification system used in Hagerman 1995a; see also Fig. 3.1).

Many combinations of absorber, energy converter and structure type are possible. For instance, the wave-absorbing movable member may be oscillating water ("oscillating water column" - OWC), an oscillating solid body or structure, or an oscillating flexible body (such as a reinforced rubber membrane separating the sea water from air, water or another hydraulic fluid). An OWC may be placed in a fixed structure or a floating structure. In the latter case, energy conversion is provided through the *relative* motion between the OWC and its containing structure. With OWCs (and also with flexible membranes separating sea water from entrapped air), the secondary energy conversion is usually realized by means of a conventional air turbine in combination with rectifying valves or a self-rectifying air turbine (e.g. a Wells turbine) without any valve. Another possibility is to use a water turbine for the secondary energy conversion as is the case, for instance, with the tapered channel device (Fig. 3.1, #1). For WECs of the type with oscillating bodies, the secondary energy conversion is more typically realised by means of hydraulic machinery. The oscillating motion is utilised to run a pump, which establishes pressure in a hydraulic fluid. Fluid drawn from the pressure reservoir runs a hydraulic motor. The reservoir and motor may be common for several WEC units.

The theory for absorption of wave energy by immersed oscillating systems has been developed from the mid-1970s onwards. An early analysis of a WEC buoy was carried out by McCormick (1974). For sinusoidal waves, it was found by Budal & Falnes (1975, 1977), and independently by Evans (1976) and Newman (1976), that for a resonant point absorber or for any heaving axisymmetrical wave-absorbing device at resonance, the maximum power that can be absorbed equals the incident wave power associated with a wave front of width one wavelength divided by 2π, if the device is placed in the open sea. Moreover, it was found by Evans (1976), and independently by Mei (1976) and Newman (1976), that a resonant two-dimensional symmetrical system oscillating in one mode (one degree of freedom) cannot absorb more than half of the incident wave energy, but that almost all incident energy can be absorbed if the two-dimensional system is sufficiently non-symmetric, as for instance the Salter (1974) Duck (Fig. 3.1, #10). All incident wave energy may then be absorbed with optimum oscillation in two modes, for instance where one mode (heave) has symmetric wave generation, while the second mode has anti-symmetric wave generation. An example is the submerged "Bristol-cylinder" device (Evans 1976) (Fig. 3.1, #12). The above specified portions of maximum absorbed power may be obtained only for unconstrained amplitudes, in which case the waves are sufficiently low to avoid the restriction that the required optimum oscillation amplitudes are limited by design specifications.

The optimization problem of maximizing the absorbed wave power is more complicated for a system of several interacting bodies than for the previously considered single body. For instance, it cannot be assumed (as in the case of one single-mode oscillating body) that resonance provides optimum phase condition. Theories have been developed for unconstrained amplitudes, both for the case of oscillating bodies (Budal 1977; Evans 1979; Falnes 1980) and for the case of OWCs (Falcão & Sarmento 1980; Evans 1982) as well as for the case of interacting OWCs and bodies (Falnes & McIver 1985; Fernandes 1985).

Optimization with constraints is even more complicated. Some preliminary studies were made in the early 1980s. A simple optimization problem under one global constraint (upper bound on the sum of all amplitudes squared) was solved mathematically by Evans (1981), whose analysis has later been extended by Pizer (1993) and by Falnes (2000). If the waves are sufficiently high, one may assume that all amplitudes have reached their design limits, and it remains to optimize the phases (Falnes & Budal 1982). At more moderate wave heights, when some of the amplitudes are still unconstrained, while other of the amplitudes have reached their design limits, numerical optimization procedures may be adopted (Thomas & Evans 1981; Kyllingstad 1982).

Because real sea waves are not sinusoidal, it may be beneficial to apply control-engineering techniques in order to optimize the oscillation for maximizing the absorbed power or the converted useful power. Optimum control of wave-energy converters was proposed in the 1970s by Budal (Budal & Falnes 1977), and independently by Salter (1979; Salter *et al.* 1976). Milgram (1970) studied the application of control engineering to a wave-absorbing paddle in a wave channel, but utilization of the absorbed wave energy was of no concern in this study. It was soon realized (Budal & Falnes 1980; Naito & Nakamura 1986) that optimum control in real sea waves (non-sinusoidal waves) requires knowledge of the wave or of the optimum oscillation some seconds into the future. This requires some form of wave prediction. Otherwise, sub-optimal strategies have been proposed (Perdigão & Sarmento 1989; Clément & Maisondieu 1994), where only present and past information is applied. After the initial studies in the 1970s in Edinburgh (Scotland) and Trondheim (Norway), theoretical and experimental studies on optimum or sub-optimum control were also been carried out in England, Japan, Portugal and France. For more details, the reader is referred to a recent review paper (Falnes 2002).

Optimum motion is more important the smaller the physical size of the WEC. One motivation for smaller size is the larger ratio between the amount of converted energy and the size of the WEC (Falnes 1994). Other advantages with moderate size are a relatively smaller development cost and the possible benefit of mass-produced wave-power plants in the future. To realize optimum control, tailored electronic software and hardware need to be developed, along with appropriate mechanical components for control and power take-off. Thus optimally controlled WECs require a more advanced technology than is the case with more conventional-technology prototype (or semi-prototype) WECs tested in the sea to date (2003).

3.2 Wave Energy Conversion Developments

Tables 3.2, 3.3 and 3.4 list wave energy convertors that are situated on (or planned for) the shoreline, the near-shore to offshore zone, and the offshore zone, respectively. Information provided in these tables includes the primary energy conversion process employed, the country that developed the system, the location of the device, and its current status, i.e. operational or advanced-stage-development (see definitions below).

- The *operational* category comprises full-scale devices, chiefly prototypes, that are currently operating (or have operated) where the energy output is utilized for the production of electricity or other purpose; also includes full-scale devices at an advanced stage of construction.

- The *advanced-stage-development* category comprises: (a) devices of various scales, including full-scale, that have been deployed and tested *in situ* for generally short periods, but where the energy output has not been utilized for the production of electricity or other purpose (in most cases plans call for such devices to be further developed and deployed as operational wave energy systems); and (b) full-scale devices planned for construction where the energy output will be utilized for the production of electricity or other purpose. Note: devices at an early stage of development are not included.

Full details of all the plants and schemes listed are provided in the country-by-country review of wave energy conversion projects and activities found in Chapters 7 through 10; information on systems being researched or in the early stages of development, and on related wave energy conversion activities, are also included in these chapters.

Category #	Energy Conversion Process	Country	Location	Site	Status
(1)	(2)		(3)	(3)	(4) (5) (6)
1.1	Fixed OWC	Australia	Port Kembla	Breakwater	Adv. Stage Dev.
1.1	"	China	Dawanshan I.	Shoreline	Operational
1.1	"	China	Shanwei		Adv. Stage Dev.
1.1	"	India	Vizhinjam	Harbour	Operational
1.1	"	Japan	Sanze	Shoreline gully	Operational
1.1	"	Japan	Sakata Port	Breakwater	Operational
1.1	"	Japan	Kujukuri-Cho	Breakwater	Operational
1.1	"	Japan	Haramachi		Operational
1.1	"	Mexico			Operational (seawater pump)
1.1	"	Norway	Toftøy	Cliff wall	Operational
1.1	"	Portugal	Pico (Azores)	Rocky gully	Operational
1.1	"	U.K.	Isle of Islay	Shoreline gully	Operational
1.1	"	U.K.	Isle of Islay	Cliff face	Operational
1.2	Reservoir filled by wave surge	Norway	Toftøy	Gully & interior bay	Operational
1.3	Pivoting flaps	Japan	Muroran Port	Seawall	Operational
1.3	"	Japan	Wakasa Bay	Seawall	Operational

Table 3.2, Shoreline wave energy conversion devices
(Key to table follows Table 3.4)

Category # (1)	Energy Conversion Process (2)	Device Name	Country (3)	Location (3)	Status (4) (5) (6)
2.1	Freely floating OWC	-	China	Various	Operational (navigation buoy)
2.1	"	-	Japan	Various	Operational (navigation buoy)
2.1	"	Kaimei Floating Platform	Japan	Yura	Operational
2.2	Fixed floating OWC	Mighty Whale	Japan	Gokasho Bay	Operational
2.2	"	Sperbuoy	U.K.	Plymouth	Adv. Stage Dev.
2.2	"	Shim Wind-Wave System	South Korea		Adv. Stage Dev.
2.3	Bottom mounted OWC	Osprey	U.K.	Thurso	Adv. Stage Dev.
2.4	Reservoir filled by wave surge	Floating Wave-power Vessel	Sweden		Adv. Stage Dev.
2.5	Flexible pressure device	SEA Clam	U.K.		Adv. Stage Dev.
2.6	Submerged buoyant absorber sea-floor RP	-	-	-	-
2.7	Heaving float in bottom-mounted or moored floating caisson	ConWEC	Norway		Adv. Stage Dev.
2.8	Floating articulated cylinder with inertial RP	Pelamis	U.K.	Shetland/Isle of Islay	Adv. Stage Dev.

Table 3.3, Near-shore to offshore wave energy conversion devices
(Key to table follows Table 3.4)

Category # (1)	Energy Conversion Process (2)	Device Name	Country (3)	Location (3)	Status (4) (5) (6)
3.1	Freely heaving float with sea-floor RP	OPT Wave Power System	USA, Australia	Portland, Australia	Adv. Stage Dev.
3.1	"	Danish heaving Buoy	Denmark	Hanstholm	Adv. Stage Dev.
3.1	"	Phase-controlled Power Buoy	Norway	Trondheim Fjord	Adv. Stage Dev.
3.1	"	DELBUOY	U.S.		Operational (reverse osmosis)
3.2	Freely heaving float with inertial RP	Hosepump	Sweden		Adv. Stage Dev.
3.2	"	IPS Buoy	Sweden		Adv. Stage Dev.
3.3	Contouring float with inertial RP	McCabe Wave Pump	Ireland	Shannon River estuary	Adv. Stage Dev.
3.3	"	Wave Energy Module	U.S.		Adv. Stage Dev.
3.4	Contouring float with sea-floor RP	Kaiyo Jack-up Rig	Japan	Iriomote Island, Okinawa	Adv. Stage Dev.
3.4	"	Contouring Raft	U.K.		Adv. Stage Dev.
3.4	"	Contouring Raft	U.S.		Adv. Stage Dev.
3.5	Pitching float with inertial reaction point	-	-	-	-
3.6	Flexible bag with spine reaction point	-	-	-	-
3.7	Submerged pulsating-volume body with sea-floor RP	Archimedes Wave Swing	The Netherlands	Viano do Castello, Portugal	Adv. Stage Dev.

Table 3.4, Offshore wave energy conversion devices

[Key to Tables 3.2, 3.3, 3.4: (1) Category numbers correspond to Table 3.1. (2) OWC - Oscillating Water Column. (3) May not be applicable to devices under development. (4) Adv. Stage Dev. = Advanced Stage Development (5) Refer to 1st paragraph of Sect. 3.2 for definitions of "Operational" and "Advanced Stage Development". (6) Unless otherwise stated, energy output of operational devices is utilized for electricity production.

Chapter 4

POWER TRANSFER SYSTEMS

Energy derived by wave action has to be converted into a power form that can be transmitted and used for local requirements. It is usually converted into electricity, which is fed into an electrical grid system. To achieve this, power transfer systems have to turn the slowly varying, oscillating forces of incoming waves into the fast, unidirectional forces required to drive generators that produce electricity. With few exceptions, this normally consists of a two-stage system: a mechanical rotary device coupled to an electrical generator.

There is a wide range of possible solutions to convert and transfer the energy. After the initial wave device, most systems consist of a mechanical interface, an electrical generator and a method of transmitting the output into the local electrical grid system. Variations of these conversion and transfer components are described below.

4.1 Mechanical Interfaces

With the exception of the first approach, direct mechanical, the methods listed below are proven techniques that have been adapted or proposed for use in this technology.

Direct mechanical

Only a few designs have been conceived using a purely mechanical power transfer; none have been tested in full-size practice. Indeed, two thorough reviews of wave power schemes and power transfer systems did not list this as an option (Hagerman 1995a; European Commission 1993). This fact is attributable to the difficulty in devising mechanical components that address the requirement of converting the oscillating, variable forces into a high-RPM unidirectional output. The large size components required to deal with such high forces would make this approach uneconomic as indicated in a 1992 UK wave energy review (Thorpe 1992).

Two approaches considered recently are:

- The Wave Rotor (Retzler 1996). This is a concept that uses the Magnus effect on a pair of contra-rotating cylinders to absorb wave power. The system has been designed theoretically

37

and tested as a small-scale model, but the potential problems with mechanical power transfer at full-scale have not yet been addressed.

- The OLAS System (Rebollo *et al*. 1995). This is an oscillating water column (OWC) system, where the power is extracted via a float on the oscillating water column, which is mechanically coupled to a generator by means of a "mechanical rectifier" and "speed multiplier". Few details of these aspects have been given. The approach has been evaluated on a small scale prototype (maximum output of ~ 15 kW) and was found to have a mechanical conversion efficiency ({power on rotating shaft ÷ input pneumatic/ hydraulic/ float/primary power} X 100%) of ~ 75%, which is higher than that of the alternative air turbines (see below). However, no considerations of the cost-effectiveness of this approach have been reported.

In light of the above, it is considered unlikely that a purely mechanical power transfer system will be included in any large-scale wave energy scheme.

Air turbines

Air turbines fulfil the requirements of a power transfer system by offering a simple means of turning the low velocities and high forces of air compressed by sea waves into the high speeds and low forces required by conventional electrical generators. In this respect the air column and turbine provide a cost-efficient approach to gearing. The most popular type of air turbine, the Wells (Fig. 4.1), addresses the problem of oscillating air flow by having symmetrical air foils with their chords lying in the plane of rotation (i.e. no pitch angle). This gives the turbine the property of being able to rotate in the same direction, regardless of the direction of air flow (self-rectification). The turbine has small losses during idling, and the blades are cheap to manufacture compared to the more complex shaped blades of other turbines.

Various configurations of Wells turbine have been proposed and tested:

- Monoplane turbines - these have a maximum efficiency ~ 60% for small-scale testing in steady air flows (Gato *et al*. 1996).

- Monoplane turbines with guide vanes - these have a maximum efficiency ~ 70% for small-scale testing in steady air flows (the increased efficiency comes from the reduction in the exit flow swirl), albeit with poorer behaviour against stall (Gato *et al*. 1993, 1996).

- Contra-rotating turbines - these have two Wells turbines placed close together with their blades rotating in opposite directions. Each behaves like a set of guide vanes for the other, producing a peak efficiency ~ 70% for small-scale testing in steady air flows but with a wider operating range than those using guide vanes alone (Gato & Curren 1997).

There are few performance measurements available of turbine efficiencies in real, oscillating air flows. Some measurements and theoretical considerations (European Commission 1993) indicate a much reduced efficiency (<50%) in this flow regime because the turbine is operating at its most efficient range for only part of each wave cycle. Efficiency measurements on Wells turbines under realistic oscillating flows is a priority area for further work.

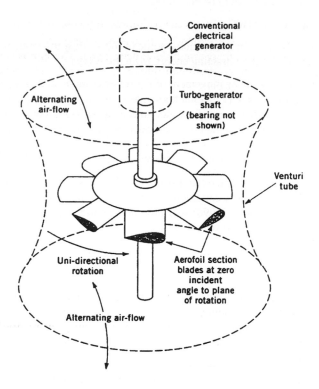

Figure 4.1 Wells turbine concept for double-acting OWC system
Source: Raghunathan, S.R. 1980. (Reprinted by permission.)

Another possibility is using a variable-pitch turbine to improve the overall efficiency (Sarmento *et al.* 1990; Gato & Falcão 1991; Salter 1994a; Taylor & Salter 1996; Caldwell & Taylor 2000). Such a turbine has been built for testing in the European Commission pilot OWC plant in the Azores (Russel & Diamantaras 1996). This refinement will add substantially to the cost of the turbine but, since the turbine cost is only a small fraction of the overall costs of the scheme (<10%), the improvements in front end efficiency more than merit the extra expenditure. Wells turbines are usually mounted on the same shaft that drives the electrical generator, which can provide a useful amount of inertial storage. In addition, decoupling a variable pitch turbine would produce a low-inertia system, which could match the power input on a wave-by-wave basis.

Other air turbines such as an impulse turbine with adjustable guide vanes have been considered (e.g. Setoguchi *et al*. 1994). The performance of such turbines appears to be superior to the Wells turbine but their cost effectiveness has yet to be determined.

The behaviour of turbines in OWC systems can be improved by use of valves:

- In high seas, valves can be used to allow some of the air flow to bypass the turbines, thereby avoiding stall while still extracting energy from the waves.

- Valves can be used to delay movement of the water column so that its velocity comes into phase with the wave force (a method known as "latching control"), which can improve the device efficiency (Justino *et al*. 1994; Falcão & Justino 1999; Korde 2002). Such a system is being developed as part of the European wave energy programme (Salter & Taylor 1996). Note that latching control is considered to be better suited for hydraulic conversion machinery than pneumatic take-off systems (see below).

Overall, the behaviour, efficiency and reliability of air turbines as used in wave energy devices have still to be determined under realistic conditions. It is expected that information from the European pilot plant (Russel & Diamantaras 1996) will provide important information in this respect but further work is required to optimize such turbines.

Water Turbines

Water turbines represent a well developed technology with minimal environmental risks (e.g., from leakage). Water from the surrounding sea provides an abundant supply of working fluid. In addition, the various designs offer some control over the volume of water flow; so that the device can cope with variations in wave power levels and so enable relatively conventional electrical generators to be used.

There are several types of turbine, designed specifically for different working pressures (the 'head') (European Commission 1994). The following is a simplified listing with some examples of their proposed or tested use in certain wave energy devices:

(a) The Pelton wheel, suitable for high-pressure operation; examples include:
- the original Bristol Cylinder (Fig. 4.2), with high pressure Pelton wheel (McAlpine 1982);
- the Hose-pump, with a high pressure Pelton wheel (Sjöström 1994) (refer to Chapter 9, Section 9.2, Sweden, for details).

(b) The Francis turbine, suitable for medium-pressure operation.

(c) The Kaplan turbine, a propeller type suitable for low-pressure operation; examples include:
- the Danish Wave Power Pump, with low pressure propeller (Nielsen *et al*. 1996) (refer to Chapter 9, Section 9.3, Denmark, for details);
- the Tapchan, with low pressure Kaplan turbine (Norwegian Royal Ministry of Petroleum and Energy 1987) (refer to Chapter 9, Section 9.1, Norway, for details).

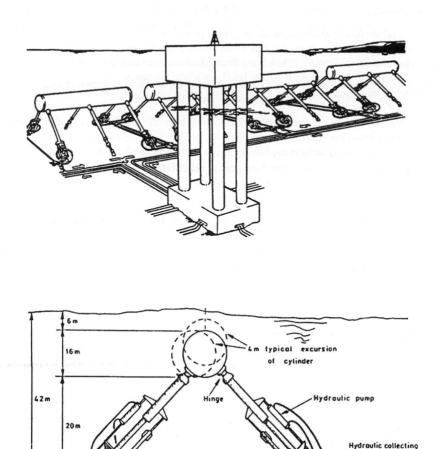

Figure 4.2 Bristol Cylinder
Source: Davies 1985. (Reprinted by permission.)

Two of these schemes (the Tapchan and the Hose-pump) have been demonstrated successfully in pilot plants, confirming that water turbines can be used in wave power schemes with little or no further development, providing well known problems are avoided (e.g., cavitation).

41

Hydraulic systems

High pressure oil systems have been used , or are proposed, for several types of wave energy device (e.g., the Japanese Pendulor, Swedish IPS Buoy, Lancaster PS Frog, Edinburgh Duck, Bristol - Cylinder, McCabe Wave Pump). They have several advantages as a power take-off technology:

- are capable of handling high power levels in a small volume;
- can adapt to different types of input motion (rotary or linear) and so be utilised on a wide range of device types;
- offer the opportunity for computer control, which enables optimization of the whole device on a wave-by-wave basis;
- can accommodate a wide range of input power levels;
- offer opportunities for significant power storage and smoothing; and
- are suitable for incorporating latching control.

Hydraulics have a proven reliability in many areas; indeed some are used on safety critical systems. However, their use in wave energy devices entails several differences to their use in conventional plant, for instance:

- Parts of the hydraulic system will be exposed to sea water. Clearly, sea water must be prevented from coming into contact with the hydraulic oil. With respect to the issue of sea water coming into contact with moving metal parts:
  – sea water can be excluded using seals (e.g. rolling "Belofram" seals for hydraulic rams);
  – sea water resistant coatings can be used; rams coated with "Ceramax" ceramic coatings have operated successfully for several years in sea water (Dijk 1992).

- The input power range could cause unacceptable loadings in hydraulic rams, i.e. large displacement under extreme conditions could bring hydraulic rams up against their end stops, resulting in high dynamic loading (only the Swedish IPS buoy has developed a means of avoiding these problems).

- Existing rotary hydraulic machinery operates at relatively low torque compared to that needed for some wave energy applications; in addition, existing machinery usually operates under fixed displacement, whereas wave energy devices operate best using variable displacement.

- Current methods of controlling hydraulic machinery can involve significant power losses; also the equipment would have problems in carrying out the complex-conjugate control to optimize the capture and conversion efficiencies of wave energy schemes (Nebel 1992).

Nevertheless, current hydraulic systems can be used for many types of wave energy devices, albeit with losses in reliability and efficiency. The design of tailor-made hydraulic systems has formed part of the European Wave Energy Programme (Russel and Diamantaras 1996), primarily at Edinburgh University (Eshan *et al.* 1996; Rampen *et al.* 1996). These systems will permit a high degree of control of a range of hydraulic devices with reduced losses. In Japan, a particular hydraulic device, a rotary vane pump, has been designed for the Pendulor system (Watabe *et al.* 1996; Osanai *et al.* 1996).

4.2 Energy Storage

Since the surface velocity of waves varies significantly with wave height and period, and power is proportional to velocity squared, it follows that wave power is subject to even greater variation. This variation can lead to an increase in both the capital costs of a wave power installation and its power losses. The provision of some form of short-term energy storage reduces both of these problems. If the scheme can store sufficient energy (i.e. corresponding to that from approximately ten wave periods), it will be able to produce almost a constant output, which would greatly facilitate its integration with the local electrical grid.

There are two main types of short-term energy storage considered for wave devices: flywheels, and pressure accumulators or elevated water reservoirs:

- Flywheels are used to an extent in some oscillating water column (OWC) devices where the inertia of the blades, generator shaft and (in certain devices) flywheel provide some storage in the form of rotational kinetic energy. However, such flywheels have incurred unacceptable energy losses from windage; so their use in all but the largest OWCs will probably be limited. Some schemes have attempted to overcome such losses by containing the flywheel in a vacuum, as well as using low-friction hydrostatic bearings, but these measures added to the costs and complexity of the system (Edinburgh-Scopa-Laing 1979). The latter approach appears feasible for shoreline or near-shore schemes, where the flywheel assembly could be mounted on shore.

- Gas accumulators, in which an inert gas is contained in a steel accumulator, are the normal method of storing energy in oil hydraulic systems. The gas is stored at a high pressure and low volume; energy is released as the gas expands to a greater volume at a lower pressure. In order to stop the gas dissolving in the hydraulic oil, a diaphragm or bladder (usually of a synthetic material) separates the two components. However, the cost of large size accumulators is prohibitively high. It is possible that costs will be lower in the future, especially if wave energy schemes create a demand for a large number of such devices. One possible alternative is to utilize large, shore-based chambers as air reservoirs for storing the energy from several devices. This has been proven in a small-scale (30 kW) scheme at Kujukuri in Japan but relatively little work has been done at a large scale (Hotta, 1996).

- Energy storage by means of water reservoirs is another method. An example is the Norwegian Tapchan plant (refer to Chapter 9, Section 9.1, Norway), where the size of the water reservoir gives a short-term storage time of approximately five minutes.

There may also be the need for longer-term energy storage, such as from nights to days or from summer to winter. This need is not a major problem in countries such as Sweden and Norway since, with limited development of wave power in these countries, the hydro-electric reservoir capacity is

43

sufficient. In Norway, where 99% of electricity production is from hydro energy, some water reservoirs are so large that they have the capacity to serve as an energy reserve not only for the following winter, but also for future years that may have significantly less precipitation than normal. If the energy system had originally been planned with, say, a 50/50 combination of hydro power and wave power, then the water reservoirs could have been made smaller since wave energy is abundant in winter, whereas the main flow of water into the reservoirs takes place when the snow melts in late spring and early summer, during a time when electricity consumption is at a low level compared to winter.

4.3 Electric Power Conversion

Electric power conversion, e.g., from alternating current to direct current, and back again, may be required in order to obtain grid-quality power from wave power devices. Various electric power conversion methods and designs have been reviewed in proposals, or tested in pilot schemes. Such techniques are generally based on previous development work in the electrical/electronic field, it being noted that advances in the power conversion field occur more frequently than in the mechanical engineering field. It is therefore advisable to consult the most up-to-date literature in this discipline, and particularly to have early discussions and agreements with the local power authority. A grid operator would not be willing to connect a power-output source that may cause disturbances or other problems on the distribution grid. Brief outlines of the possible alternatives are presented in Claeson *et al.* (1987) and SEASUN (1988) who have reviewed the subject of electrical power conversion associated with wave power systems. Beattie *et al.* (2000) discuss the production of acceptable electrical supply quality from a wave-power station.

Generators

Electrical generators convert the mechanical power at the generator shaft to electrical power. In order to achieve a high conversion efficiency from wave energy to electrical energy, without a primary energy storage system (see above), large rotation-speed RPM differences must be allowed for in the turbine-generator system. Among the various approaches is the Kvaerner multi-resonant OWC system (refer to Chapter 9, Section 9.1, Norway). There are special generators that can meet the required specifications, for example multi-pole generators for step-wise RPM changes, linear generators, and generators using reluctance changes. Linear generators, in particular, are suitable for direct connection to the mechanical reciprocating motion of a wave energy device. This approach is an example of the thrust of new research and developments, particularly in magnetic materials, that make the utilization of linear generators more feasible than previously. Other methods, for example the TAPCHAN (refer to Chapter 9, Section 9.1, Norway), have energy stores large enough to allow the generator to work at a constant RPM and at almost constant power, a requirement for connection to the distribution grid.

Power conversion at constant or nearly constant RPM

The requirement of low maintenance for sea-based units means that only alternating current motors without slip rings are possible. These can be further divided into two main types, synchronous generators (SG) and asynchronous generators (AG). In the directly connected SG the power conversion takes place within a large power range at constant (synchronous) RPM, depending on pole number and grid frequency. Outside of the power range the SG and AG fall out of phase or operation. The power direction (generator or motor action) depends on the direction of the electromagnetic torque (i.e., with or against the direction of rotation).

Electronic power conversion

Using high-power electronics (e.g., various thyristor or insulated-gate bipolar transistor {IGBT} connections) the electrical power can be converted between alternating current (AC) and direct current (DC). AC systems of different frequencies can either be connected with a direct converter (AC/AC) or pass via a DC interconnection (AC/DC/AC).

Power conversion at variable, or free, generator RPM

In some cases it is preferable to allow large variations in RPM. This is possible through power electronics, for example by employing a DC interconnection between the generator and the power grid. Due to the oscillatory wave movement, a direct linear synchronous generator for forwards- and backwards-movement is a possibility. A generator for such low speeds or RPM is heavy and, as a consequence, expensive (the weight of a 20kW linear synchronous generator is estimated to be 800 kg). The voltage produced by such a generator will be frequency- and amplitude-modulated by the random wave field and cannot be directly connected to the grid. If the DC voltages from several generators running at random RPM are added in series, the voltage variations are lessened in the DC interconnection.

4.4 Transmission to Grid

Power collecting systems connect the converters to a land-based grid. The distance between the converters and the grid is often great, which creates transmission losses. Power transmission can be accomplished in different ways. In the power plant itself, hydraulic and electrical systems are utilized. The transmission to the land-based grid generally employs sea-floor cables, but if the distance is not too great, hydraulic transfer may be possible.

Electrical transmission

Claeson *et al*. (1987) and SEASUN (1988) review the subject of electrical transmission to the grid.

Scott (2001) reviews the grid connection of large-scale wave energy projects. Thorpe (1992), Vol. 2 describes transmission systems associated with specific U.K. devices including the NEL Oscillating Water Column, the Edinburgh Duck, and the Bristol Cylinder. The advantages of high-voltage electrical transmission are the small losses, the high reliability, and relatively straightforward management. The transmission is conducted via sub-sea cables. Today there exists a vast experience of cable laying at sea, and specially designed vessels are used that can transport the entire cable (< 130 km) in one piece on a carousel drum. The power can be transmitted as AC or DC. However, with AC transmission a reactive component of the current is created, depending on the cable length and degree of insulation. This current results in additional heat losses and reduces the capacity of the line to transmit active power. At distances longer than 30 - 70 km, AC transmission results in losses that begin to be significant. For long distances, DC transmission can be used with the associated additional costs of AC/DC and DC/AC transformation. The sea water itself can be used as a conductor in AC transmission systems, reducing cable costs; but the risk for leakage currents and galvanic corrosion must be taken into account.

Electrical transformers are needed to transfer the power between grids of different voltage levels. For instance, in Sweden, distribution grids use 400 V, local grids 4 or 10 kV, and regional grids 50 kV and up, while in the UK the corresponding voltages are 440 V, 3.3 kV, 125 kV and up. The design of the electrical transmission system is a techno-economical problem of optimization. The trade-offs to be evaluated include AC vs DC power and high vs. low voltage levels (high voltage giving smaller losses, but introducing insulation problems and mechanical problems due to the thicker cables).

The experience of wind power systems, where the variable operating conditions can result in power quality problems, is of potential interest to wave power generation/transmission systems, particularly onshore and near-shore developments. As an example, the international company ABB has developed a power transmission technology that is suitable for small-scale power generation/transmission applications such as those involving wind farms. Specifically the system extends the economical power range of HVDC transmission down to a few megawatts. The technology, know as "HVDC Light", has been installed in a number of locations, including several where its capabilities have been utilized to overcome the power quality problems of wind power plants; one such example being on the Swedish island of Gotland (ABB Ltd. 2002).

Hydraulic transmission

In some wave energy devices, the secondary energy transformation is realized by pumping water or hydraulic oil. The pressure and flow of the fluid usually drives a turbine or hydraulic motor. Hydraulic oil is normally only used internally in a wave power device, while water can be pumped long distances in open or closed systems. The Swedish hose-pump concept is an example using sea water in an open system. Losses are normally proportional to velocity squared, which motivates low flow velocities combined with high-pressure or large-diameter pipes.

Chapter 5

ECONOMICS OF WAVE POWER

Despite considerable technical progress during the past twenty years in developing means of harnessing wave energy, the commercial deployment of such schemes is only just beginning. Therefore, there is a lack of experience enabling accurate costing of such schemes and, hence, an appreciation of their economic competitiveness. This chapter identifies the important factors to be taken into account in evaluating costs, and suggests an approach for determining capital and operating costs in the absence of in-service experience. It examines the likely electrical generating costs of a number of devices, including combined wind/wave-powered systems, and considers the economic prospects of wave power. Information on the economics of wave-powered desalination is also included.

5.1 Important Factors in Evaluating the Costs of Wave Energy

All too often, ocean energy developers simply state a cost of energy in cents per kilowatt-hour. The method for calculating the energy cost is usually not identified and critical financial assumptions, such as discount rates, rates of return and debt-to-equity ratio, are often not stated. Most importantly, there is usually no economic evaluation of competing technologies, so that a potential lender or investor has no appropriate benchmark to determine whether a project is commercially viable.

The most accurate method of addressing the above factors is via a formula based on the levelized revenue requirement methodology published by the U.S. Electric Power Research Institute in its Technical Assessment Guide, EPRI TAG™ (Electric Power Research Institute 1987, 1993; see also Hagerman 1995b). The cost of energy is computed by levelling a power plant's annual revenue requirement over the service life of the plant and dividing it by the plant's annual output. If the energy would be sold at this price, the total collected revenue would have the same present value as the sum of all fixed charges and expenses paid out during the plant's life, allowing for discounting of future costs and revenues. This makes it possible to compare alternative designs or technologies in terms of a single index -- the "levelized" cost of energy.

47

A project's annual revenue requirement can be divided into two categories: (a) fixed charges; and (b) variable costs or expenses. Fixed charges are annual payments associated with making a capital investment and are "fixed" because they are determined at the time an investment is made, regardless of how much the investment is subsequently used. They are financial obligations associated with building the plant, and include the costs of all aspects of the scheme, including the civil structure, mechanical and electrical plant installation, electrical transmission to shore and (possibly) the subsequent link to the nearest grid connection point. The provision of adequate spares and payment of rates (local taxes) could also be assigned to these costs. Expenses, on the other hand, cover the annual cost of using an investment. They are annual payments associated with plant operation and maintenance (O&M), and include fixed O&M, variable O&M, and (where relevant) fuel. Fixed O&M expenses are incurred on a regular basis, and do not depend on how much electricity the plant generates. Salaries, insurance and the lease of land are examples of fixed O&M expenses. Variable O&M expenses are linked directly to the amount of energy produced and include consumable items such as filters and lubricants, which are replaced after a certain period of operation. Fuel expenses also depend directly on the amount of energy produced, as well as on the thermal efficiency of the plant. The sale of electricity and any co-products, e.g. fresh water, is credited towards expenses. If product sales exceed expenses, then the difference represents profit and is subject to tax. It is important to note that all these costs are scheme and site specific.

5.2 Determining the Capital and Operating Costs of Wave Energy Schemes

The capital costs for wave energy include those monies associated with construction, assembly and installation of a wave power scheme. To date, no large-scale offshore wave power scheme has been built. Existing schemes have been prototypes with all the additional costs attendant on devices at such a stage of development. Therefore, there is no mature wave energy industry which could be used as the source of cost data. As a consequence, the capital costs associated with wave energy have to be evaluated using data from other relevant areas, such as the offshore oil and gas industry and large-scale civil and mechanical engineering.

At the simplest level, an approximate estimate of costs can be obtained using average figures from similar developments, making adjustments to account for any differences (costing by analogy). These adjustments could be made either on the basis of relative sizes, complexity etc. or a combination of characteristics. This approach is most suited to estimating the costs of those schemes with only outline designs and for which similar schemes with known costs exist. The lack of any structures that are closely analogous to wave-energy schemes restricts the applicability of this approach within the current review.

At the other extreme of sophistication, a more definitive, conventional (or "bottom-up") costing can be made. This requires the establishment of detailed drawings and construction plans, which are used to establish a complete work breakdown structure, with unit rates etc. being developed for each cost category. The early stage of development of wave energy would make such a costing method inappropriate for general use, although aspects of some of the devices, which use relatively conventional technology and construction methods, could be costed by this approach. In addition, it would be difficult to ensure compatibility between "bottom-up" costs for different devices as produced by different estimators.

An intermediate method is that of obtaining a budget estimate based on parametric costings. In this process, costs are calculated using a functional relationship between some characteristic of an item (e.g. weight or power output) and its cost. Such relationships are derived either from past experiences or through engineering judgement. In many cases such relationships are similar to those used as base rates in "bottom-up" costings (e.g. cost per ton of concrete). This approach requires outline drawings and specifications together with unit rates for materials, labour, plant, transport etc. Such information is generally available for all but the least defined wave energy devices, although engineering judgement might have to be used to determine some aspects of a feasible power generation scheme. Therefore, it would appear that, in general, this level of costing would be the most suitable for wave energy. Such an approach has been developed as part of the UK and European wave energy programs (Thorpe 1992; European Commission 1993).

5.3 Generating Costs of Wave Energy Schemes

Early proposed wave energy schemes were evaluated using the parametric approach and found to be too expensive, with capital costs of >$3,000/kW. However, there has been a steady improvement in the capital costs (in $/kW), and O&M costs (in ¢/kWh), of both specific devices (e.g., the Duck and Clam) and generic devices (e.g., OWCs) with time; so that there are now several systems with expected capital costs and performance for the large-scale deployment of mature devices of ~$1,500/kW. Table 5.1 illustrates the predicted costs and performance for generic types of wave energy device

Cost & Performance	Shoreline	Near-shore	Offshore
Unit Costs ($/kW)	1,800-2,100	1,500-1,800	1,500-3,000
O&M & Insurance Costs ($/kW/Year)	30-45	42-48	30-90
Availability (%)	94-96	93-96	90-95
Annual Output (kWh/kW)	2,000-2,500	2,200-2,500	3,000-4,000

Table 5.1 Cost and performance characteristics of generic wave energy devices
Source: Thorpe 2000. (Reprinted by permission.)

Figures 5.1 and 5.2 show, respectively, independently predicted electricity generating costs (in UKp/kWh) of near-shore (including onshore) and offshore devices against the year in which the device was designed (Thorpe 1998). These costs, which are site specific and in respect of representative wave climates, show reductions to approximately 5 p/kWh (7 ¢/kWh) in 2000 for both categories (at 8% discount rate over the lifetime of the scheme). There are also several other schemes that claim to be able to produce electricity at similar costs to those presented in Figs. 5.1 and 5.2, indicating that, with a suitable wave climate, generating costs of 3.5-8 ¢/kWh should be achievable (e.g., Sjöström 1994; Sjöström *et al.* 1996; Thorpe 1997). However, these schemes have either not been evaluated independently or else are in the early R&D stage and, therefore, their costs and performance are subject to considerable uncertainty.

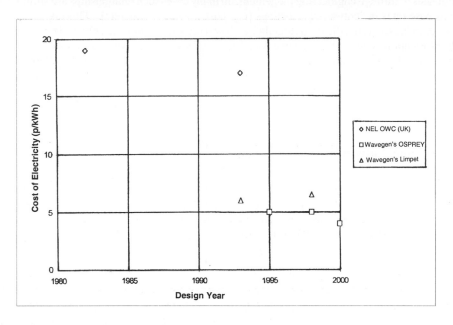

Figure 5.1 Evolution of electricity costs for onshore and near-shore devices
Source: T.W. Thorpe, Energetech, Australia. (Reprinted by permission.)

The economics of wave energy have shown a gradual improvement with time. This improvement is similar to that for other new technologies, where developing more understanding results in reduced capital costs (Figure 5.3). Falnes (1996) shows that a certain amount of funding of research and development is required to bring wave energy converters to a commercial level and that when this level has been achieved there is a potential for selling wave-energy converters in a huge market. However, in comparison with most other renewables, wave energy has achieved such improvements despite having received little financial support in recent years. Several radically different designs have recently been tested and taken through to the prototype stage, while some are near the stage of proceeding to their first commercial scheme. To achieve this, the designers have had to improve

50

their concepts, thereby reducing the generating costs from those originally estimated to values of ~ 7 ¢/kWh.

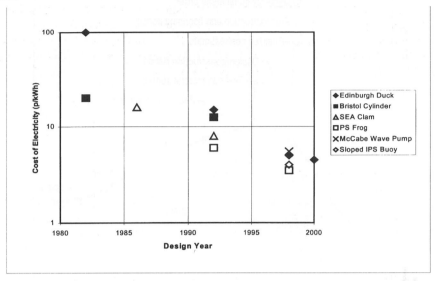

Figure 5.2 Evolution of electricity costs for offshore devices
Source: T.W. Thorpe, Energetech, Australia. (Reprinted by permission.)

In addition to pure wave energy schemes, there are novel approaches that utilize the wave energy device as a platform for mounting wind turbines (Shim 1996; Thorpe 1995, 1997). These devices not only generate electricity at a low cost (6-9 ¢/kWh) but also have the added advantage that separate output from the wind and wave schemes offer greater "firm power" than either alone. However, it should be emphasized that in the above description of the economics of wave energy, the predicted costs are for mature devices following successful R&D and deployed in multi-device schemes. The generating costs of the first, individual devices will be much higher because of:
- - technical immaturity (learning curve benefits will follow);
- - perceived risk (which will inflate the costs of the initial schemes);
- - large mobilization and demobilization costs which are loaded onto the cost of a single device (in future these would be defrayed across a number of devices); and
- - lack of economies of scale initially.
As a result, the generating costs of the initial devices can be a factor of two or three times higher than the costs indicated above.

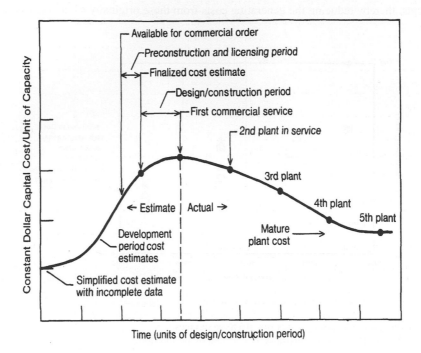

Figure 5.3 Capital cost learning curve
Source: G. Hagerman, Virginia Tech. Research Institute, Alexandria, VA, USA. (Reprinted by permission.)

5.4 Comparison of Economics of Electricity Generation

Comparison with other renewable energy generation costs

The costs of electricity from renewable energy sources depend upon many factors (e.g. the type of source, the efficiency of the plant, its location). Therefore, it is impossible to give definitive costs for "electricity generation" unless a particular plant is specified. Table 5.2, which is based in part on a global review presented at the 1993 World Energy Congress, presents representative future costs of electricity from various renewable sources. The table shows that combined wind-wave installations deployed in energetic seas off the UK would be economically competitive with the future costs of solar thermal and photovoltaic (PV) plants, and probably competitive with onshore wind installations.

System	Location	Date	Cost (¢/kWh)
Solar thermal; parabolic trough	New Mexico, USA	2020	7.5-11
Solar thermal; parabolic dish	New Mexico, USA	2020	6-10
Solar, thermal; central receiver	New Mexico, USA	2020	5-9
Photovoltaic	New Mexico, USA	2020	5-14
Photovoltaic; thin film	New Mexico, USA	2020	6-10
Photovoltaic; multiple thin film	New Mexico, USA	2020	4-7
Wind turbine (6-9 m/s wind spd.)	–	1995	3.6-6.5
Wind turbine (6-9 m/s wind spd.)	–	2000	3-5.5
Wind turbine (6-9 m/s wind spd.)	–	2010-2020	2-4.5
Combined wind-wave system	South Korea	1995	11-18
Combined wind-wave system	UK	1995-1999	6-9

Table 5.2, Representative costs from renewable energy sources
Source: Based on World Energy Congress 1993; Cho & Shim 1999; Thorpe 1995.

Comparison with conventional generation costs
The costs of electricity from conventional fossil-fuel fired power stations depends on the country, the fuel type, the plant efficiency, the manner in which the plant is used (e.g., base load, load following), and the cost of any abatement technologies. Therefore, it is impossible to give definitive costs for "electricity generation" unless a particular plant is specified. However, indicative generation costs of typical plants have been derived for comparison with electricity from renewable sources (Table 5.3).

Comparison with electricity prices
There is another potential way of evaluating the competitiveness of a wave energy scheme, which entails comparing the costs of wave-powered electrical generation against the price that customers pay for electricity. This method of cost comparison is becoming increasingly relevant for large-scale industrial companies (e.g., oil refineries, chemical plants) where such companies often find it economical to build and operate their own generating stations, usually combined-cycle gas turbine (CCGT) plants. While it is unlikely that, from this perspective, large industrial users would invest in wave energy schemes (they need a regular supply of electricity), it is possible that local communities would invest in a small local scheme. These could be communities that decide to

invest in "green power", or else communities that are isolated from the electricity grid (e.g. on small islands). In the latter case, wave energy would be competing against diesel generation. An indicative summary of the price of electricity in selected major industrialized countries is shown in Table 5.4, on which basis it will be noted that electricity from wave energy would be competitive with electricity purchased by households. By way of comparison, Table 5.5 shows the prices in small island countries in the South Pacific where diesel generators are used for electricity production.

Station Load	Date	Cost (¢/kWh)
Base load	1995-2020	5
Intermediate load	1995	6-7
Intermediate load	2020	6-7

Table 5.3, Electricity generating costs for conventional fossil-fuel fired power stations
Source: World Energy Congress 1989.

Country & date	Reference	Price category	Cost (¢/kWh)
USA – 1994	IEA, 1994a	Electricity price to industry	4.7
USA – 1994	IEA, 1994a	Electricity price to households	8.4
Japan – 1993	IEA, 1994a	Electricity price to households	14 - 24
UK – 1995	DTI, 1996	Electricity price to industry	6.4
UK – 1995	DTI, 1996	Electricity price to households	8 - 13.4

Table 5.4, Examples of electricity prices in major industrialized countries
Source: As indicated in table.

Country & date	Population	Price category	Cost (¢/kWh)
Fiji Islands – 2002	800000	Electricity price to households	10.54
Niue – 2002	2000	Electricity price to households	12.7
Solomon Islands – 2002	400000	Electricity price to households	10

Table 5.5, Examples of electricity prices in small island countries (South Pacific)
Source: South Pacific Applied Geoscience Commission. (Reprinted by permission.)

External costs
Conventional methods of electricity generation can cause environmental damage to a range of receptors, including human health and natural ecosystems. Such damages are referred to as 'external costs' or 'externalities', as they are not reflected in the market price of energy. Traditional economic assessment has ignored externalities, but there is growing interest in adopting a more sophisticated approach involving the quantification of the environmental and health impacts of energy use and their related external costs. This is driven by factors such as the need to integrate environmental concerns when considering different fuels and energy technologies and the increased attention to the use of economic instruments for environmental policy. As a result, there has been an increasing recognition of the importance of external costs and the need to incorporate them in policy and decision making. This led the European Commission to fund a major R&D Research Programme (known as the ExternE Project) to attempt to use a consistent 'bottom-up' approach to evaluate external costs associated with energy use (European Commission 1995a).

The results of the first phase of the project (Table 5.6) show indicative external costs from the fossil-fuel generating technologies. In addition to these costs, there is also the contribution of fossil-fuel consumption to global warming. The ExternE study has calculated values for potential external costs due to the latter, based on a review of estimates in the literature. These are shown in Table 5.7, although it should be stressed that there is considerable uncertainty attached to the costs (for details pertaining to the individual fuels refer to European Commission 1995a, 1995b). The ExternE Study has also calculated the impacts from renewable energy technologies. These external costs were shown to be considerably lower than those from fossil fuel stations, confirming the environmental benefits of renewables in replacing current generating technologies.

If future environmental penalties are included in the costs of electricity from fossil-fuel fired plants (based on initiatives such as the ExternE Project), this would increase the cost of conventional generation. On this basis, the cost of electricity from wave energy devices in suitable locations could be economically competitive with conventional generation plant.

5.5 Economics of Wave-powered Desalination

McCormick (2001) reports on the economics of a wave-powered pumping system designed to deliver high-pressure, pre-filtered salt water to a reverse-osmosis (RO) desalination unit. The system, called the McCabe Wave Pump (MWP), which has been designed and deployed as a prototype in the Shannon Estuary, Ireland, is self-contained and, therefore, is suitable for remote locations (refer to Chapter 9, Section 9.8, Ireland, for details of the MWP). Not including the desalination plant, the stated cost of supply water is about $0.95 per 1000 U.S. gallons. When the cost of the RO-desalination system is included, the delivered cost of potable water is approximately $7 per 1000 U.S. gallons.

Receptor	Coal (¢/kWh)	Oil (¢/kWh)	Gas (¢/kWh)
Public health	1.28	1.41	0.27
Occupational health	0.2	0.06	0.01
Crops	0	0.01	0
Timber	0	0.01	0
Marine ecosystems	0	0.03	0
Materials	0.06	0.09	0.01
Noise	0.03	0.03	0
Totals	1.57	1.64	0.29

Table 5.6 Indicative estimates of external costs (potential damages) for fossil-fuel life cycles
Source: European Commission 1995a, 1995b

Reference	Coal (¢/kWh)	Oil (¢/kWh)	Gas (¢/kWh)
Cline, 1992 (a)	1.92	1.28	0.77
Frankhauser, 1993 (b)	1.28	0.77	0.51
Tol, 1995 (c)	2.30	1.54	1.02

Table 5.7 Examples of indicative external costs for global warming
Source: T.W. Thorpe, Energetech, Australia. (Based on references in table.) (Reprinted by permission.)

[Key: (a) Cline W.R. 1992 *The Economics of Global Warming*. Washington DC: Institute for International Economics; (b) Frankhauser, S. 1993 Global warming damage costs: some monetary estimates. *CSERGE Working Paper GEC 92-29*. London, UK: University College London; (c) Tol, R.S.J. 1995 The damage costs of climate change – towards more comprehensive calculations. *Environmental & Resource Economics* **5**, 353-374. Note that the costs of Cline and Frankhauser are based on a 0% discount rate; those of Tol are based on a 1% discount rate.]

Chapter 6

ENVIRONMENTAL AND SOCIOECONOMIC IMPLICATIONS

All production (transformation), transport and uses of energy affect the environment and will, to a greater or lesser extent, be in conflict with other user interests or change the conditions for these interests. This applies to the utilization of the indigenous energy source of ocean waves. As with other energy sources, wave power has both environmental and socioeconomic implications; and these aspects must be considered when research, development and planning are carried out. The main environmental benefit of wave energy schemes is that during their operation they produce electricity without the emissions that are associated with conventional fossil fuel power stations. It follows that there would be a net benefit for society if "clean" wave energy were to replace conventional, polluting energy sources. Wave power, however, is not without some negative environmental impacts, although none of these are considered to be critical. Table 6.1 provides a general summary of the most common environmental impacts, both positive and negative, of wave energy devices.

6.1 Environmental Benefits

During their operation, wave energy devices produce none of the atmospheric pollutants commonly associated with conventional electricity generation (e.g. carbon dioxide, nitrogen oxides and sulphur dioxide). However, there are emissions of these gases during infrastructure development, the most significant being associated with the processing of materials and the manufacture of components (International Energy Agency 1998). These will vary depending on the type of device, its location and its country of manufacture. In this chapter, life cycle emissions have been evaluated for a representative wave energy device, Wavegen's Osprey (Wavegen 2000; Thorpe 1995). The information has been derived for a site-specific scheme, and is therefore only indicative. The results show emissions that are an order of magnitude lower than those from coal fired stations (Table 6.2), indicating that wave energy can make a significant contribution towards the reduction of emissions associated with electricity generation.

Issue/Aspect	Impact(s)
Sedimentary flow patterns	Changes in sediment transport and deposition may occur due to the siting and operation of wave power plants; flow patterns are difficult to predict and will require site-specific studies.
Coastal erosion	Near-shore schemes may affect coastal erosion by modifying wave activity: usually a beneficial effect.
Construction/maintenance sites	Possible visual and acoustic impacts; existing construction facilities should be used where possible.
Device moorings	Potential for impacts, e.g. on shipping, if moorings break away; careful design minimizes this.
Acoustic noise	Noise from onshore and near-shore installations may annoy some persons in nearby dwellings; and also may affect livestock, etc. Noise emissions, including those from offshore installations, will be directed mainly into the atmosphere and hence are unlikely to have major impacts on whales, etc.
Recreation	Visual impacts of large-scale shoreline projects may impact tourism; however, offshore schemes could enhance water sports by providing sheltered waters
Hazards to navigation	Easily minimized using conventional technology; note that project lights of wave power plants could be an aid to navigation.
Marine biota	While construction of wave power plants may have some initial adverse effects, installations can have a positive impact over the long term by serving as artificial reefs
Fishing	Exclusion zones around offshore devices could impact fishing operations, but they may prove beneficial to the resource; near-shore installations can assist mariculture operations through creation of sheltered conditions.
Endangered species	Marine mammals may be vulnerable, e.g. structures may act as a barrier; possible impacts avoided by not siting in habitats important to such species.

Table 6.1, Possible environmental impacts of wave energy devices
Source: Working Group Members.

Wave energy can have a number of other benefits, both environmental and social. For example, in remote coastal areas and small islands it can help to reduce the reliance on auxiliary (diesel) power stations. In addition to the resultant reduction of the emission of combustion gases to the atmosphere (see Figure 6.2), the transport of the fuel to the site, often by ship, is largely eliminated. This in turn reduces the environmental risks inherent to this means of transportation. Many remote coastal areas receive their electricity via overhead transmission lines, which are sometimes perceived as creating adverse visual impacts. These kinds of aesthetic concerns would be reduced by having separate wave power installations serving individual coastal communities.

Most offshore structures attract fish in large numbers through a "reefing" effect. Their surfaces foster growth of attached organisms and these, in turn, attract larger organisms that graze on attached life, or which prey on associated organisms (Golomb 1993; Thorpe & Picken 1993). The increase in marine biological productivity from such a reef is small; nevertheless, it creates a new habitat that would not otherwise exist within the local ecosystem. Evidence indicates that the reefing effect does not produce a significant increase in fish production but serves to aggregate local populations of fish (Side 1992). This aggregation effect appears to be backed up by operational experience - studies at an offshore wind-farm have shown that codfish numbers increased around the foundations of the wind-power generators (Schleisner & Nielsen 1997).

6.2 Environmental Impacts

Coastal and near-shore areas are generally the most diverse and productive of all ocean environments. Wave-induced impacts can have important roles in maintaining distinctive physical environments and that provide habitat for the establishment of biological communities (Bascom 1980; Ippen 1966). However, physical alteration of the environment, e.g., during construction of coastal and offshore installations, can have negative impacts on marine biodiversity (Vogel 1981; Norse 1993).

Wave energy conversion, like all energy sources, is associated with a range of possible environmental impacts. In general, these can be minimized, and sometimes avoided completely, by employing best-practice procedures. Potential impacts and appropriate methods of amelioration are discussed below. Since wave energy power plants are still at either an experimental or early deployment stage, there is little direct operational experience; so this discussion of environmental consequences needs to be reviewed after the installation of the first schemes.

Gas	Wave Energy Plant Emissions (g/kWh)	Coal Fired Plant* Emissions (g/kWh)
CO_2	24.6	955
SO_2	0.24	11.8
NO_x	0.1	4.34

Table 6.2, Life cycle emissions from a wave energy device in relation to a coal-fired plant (*– pulverized fuel plus flue gas desulphurization.)
Source: T. Thorpe, Energetech, Australia. (Reprinted by permission.)

Loss of amenity

Shoreline and near-shore wave energy developments can usually be observed from a large distance unless obscured by the shoreline topography or atmospheric conditions. Near shipping routes, safety requirements will necessitate the use of navigation lights and high contrast colours for the benefit of sea traffic. Associated onshore transmission schemes (transformer stations, overhead lines, etc.) will also have a visual impact. Therefore, all but the smallest installations may prove to be environmentally unacceptable along coastlines that feature important aesthetic values (e.g. wilderness areas or tourist spots). The UK Isle of Islay device has shown that a single, isolated device in a tourist area can become an attraction for visitors, but the reverse is likely to be true in the case of larger-scale (e.g., multiple) deployments. Offshore installations will have less visual impact, especially if their onshore transmission networks are buried (which is expensive) or re-routed to avoid visually important areas.

Another potential concern is noise. A Wells turbine in an onshore or near-shore oscillating water column (OWC) device can emit uncomfortable levels of noise (in effect, it can act as a siren). Experience has shown that this noise can be reduced to acceptable levels (or possibly eliminated altogether) by careful design and/or through acoustic muffling techniques.

Coastal deposition and erosion

Waves and currents have an important effect on the movement of small solid objects, in particular sand, on the sea bed and at the shoreline. This "littoral drift" process results in the erosion of shorelines at some locations and the building up of new shore-fronts at others. Man-made structures, such as jetties and groins, have been used in attempts to control this drift. They typically extend across the surf zone, thereby impeding the wave stress or agitation and protecting important tourist areas, such as beaches, from erosion. Depending on their type, size and location, wave energy installations can affect these coastal transport processes as follows:

60

- Floating devices will have a unique impact on fixed, near-shore structures because any wave energy that is not absorbed by the floating device will pass by it and would therefore still be available to energize the littoral drift process.

- Caisson-based devices have a higher potential impact than floating devices because the caissons will reflect any incident wave energy that is not absorbed, and because they would typically be sited closer to shore; under which circumstances a distinct, low-energy "shadow zone" could be created behind a row of such devices.

In view of these impacts, care must be exercised in selecting the location of wave power plants, especially in the case of near-shore caisson devices. It is recommended that the environmental effects of large-scale installations should be model-tested as part of pre-project activities.

Impacts on ecosystems

The installation of wave energy devices, particularly onshore and near-shore designs, could have impacts on fish since construction activity will disturb the sea-bed. In particular, such activity will likely affect benthic species, where disturbances could alter the composition of the community for a period of time, and might also reduce the transparency of the water, thereby influencing other local flora and fauna. However, the impacts are generally not likely to persist for longer than one season providing that ecologically sensitive areas, such as breeding grounds for fish and other marine species, as well as commercial shellfish growing areas, are avoided. In addition to potential impacts resulting from the construction of the plant itself, impacts could result from the laying of electrical transmission cables. Some guide to the potential level of impact of cable laying can be gained from the experience of the offshore oil industry in laying oil and gas pipelines. The laying of such pipelines causes a disturbance corridor of about five meters in width (European Commission 1995b), though effects due to suspended sediment levels from associated dredging operations may affect organisms across a wider area. However, after completion of the pipeline installation the area is usually re-colonized. Consequently, little long-term damage is expected from cable laying on the sea-floor.

As indicated in Table 6.1, there may be potential environmental consequences for marine mammals as well as for land mammals. Many species of marine mammals, and also of seabirds, are highly migratory, and therefore may be impacted by wave energy schemes. Seabirds that commonly nest on offshore rocks and stacks may attempt to colonize caisson-based wave energy devices, since these are designed to be unmanned with only occasional service visits.

Natural resources such as beaches, coastal wetlands, etc. will likely be impacted where power cables from offshore wave energy installations come ashore. In particular, dune vegetation and inter-tidal zones represent ecologically sensitive areas that are at potential risk from such projects.

Impacts on fishing

Wave energy installations take up a small area of the sea that typically becomes excluded from other uses such as fishing. This area comprises not only the general "footprint" of these devices, such as their foundations and moorings, but also a small safety "exclusion" zone around the devices. In addition, there are areas of the sea bed adjacent to any sub-sea transmission cables that might also need to be designated off-limits to commercial fishing because of the possibility of damage to cables by bottom fishing gear (cables laid in trenches with sufficient protective cover would be safe from such disturbance). The establishment of areas in which fishing is restricted may actually be beneficial to the fishing industry since there has been increasing interest recently in designating marine protected areas, i.e., areas that are closed to commercial fishing for the purpose of conservation management aimed at revitalizing fish populations (Shackell & Willison 1995).

It is possible that wave energy schemes might be excluded from important fishing areas but there should still be sufficient coastal regions available to exploit wave energy resources. In practice, the sub-sea parts of wave energy installations might provide a beneficial refuge for fish (fishing is often better over sunken wrecks) as well as new fish habitat, thereby increasing the abundance of fish in the vicinity of an installation (refer to Section 6.1, above). Overall, wave energy development is not expected to have significant adverse effect on fishing if sensitive coastal areas are avoided. Nevertheless, discussions with fishermen and other concerned parties should form part of the consultative process prior to the implementation of any wave energy scheme.

There is also the potential for impacts on fish farming operations in near-shore areas. It is important that the interactions between the two industries be recognized, so that these different uses of the marine environment can be accommodated and managed in harmony.

Impacts on shipping

The planners of wave-power schemes need to take into consideration shipping traffic levels, both current and potential future, when a decision is being made on a specific location. In particular, plants should not be situated in the vicinity of shipping lanes, harbour entrances, near pilot stations, or in coastal waters where sea traffic is frequent. The imposition of safety zones, as well as the use of navigation lights and radar reflectors, will minimize the risk of collision. However, there could be specific problems for wave-power devices that have a small freeboard, such as the Hosepump, and special consideration will have to be given to mechanisms aimed at preventing collisions with such designs. The adequacy of moorings should be a prerequisite to obtaining the necessary insurance to deploy floating devices; if such a device were to break away from its moorings it would likely constitute a major hazard to local shipping.

Impacts on recreation and tourism

The most important potential influence of wave energy developments on recreation and tourism pertains to visual impact. This applies primarily to onshore and near-shore installations, and in such cases could prove an obstacle to large-scale deployment of wave energy schemes in areas that generate tourism or have aesthetic importance. These aside, the impact of such installations on recreation will be local and site-specific. Sheltered waters behind wave energy devices could prove attractive for some sports such as scuba diving and kayaking. On the other hand, wave-energy schemes could reduce the size of areas suitable for wind surfing and some types of sailing. Again, prior discussions with interested parties should be undertaken to assess the implications of such conflicting uses.

Marine pollution and related concerns

Emissions from wave-energy installations might occur as a result of bad practice or accidents, e.g., schemes that utilize oil-hydraulic systems could spill oil if the hydraulic circuits are breached. Fail-safe systems should prevent the oil from reaching the surrounding water; in some schemes, sea water or fresh water could be used instead of oil. Spillages and sewage discharge from construction vessels are possible. Some developments might require the use of antifouling agents, many of which are toxic to aquatic species. Environmentally safe options for antifouling coatings exist, and toxic antifouling agents can be employed safely, e.g., incorporated into a rubber coating applied to the affected areas.

Unlike conventional fossil-fuel technologies, wave energy installations produce no greenhouse gases or other atmospheric pollutants while generating electricity. However, emissions do arise from other stages in their life cycle (i.e., during the chain of processes required to manufacture, transport, construct and install the wave energy plant and transmission equipment – see also Section 6.1). Emissions from these pre-operational stages have been evaluated to produce a fair comparison with emissions from fossil fuel based generation (Thorpe 1999b). Fig. 6.1 shows the relative emissions of the greenhouse gas carbon dioxide from a representative near-shore wave energy device (oscillating water column) and the following conventional generating technologies: combined cycle gas turbines (CCGT); modern coal-fired plant (pulverized fuel with flue gas desulphurization); and a mix of UK generating plants (coal, oil, large-scale hydro, CCGT and nuclear) (Bates 1995). Clearly, wave energy offers considerable scope for reduced emissions of this gas; and similar benefits are found for other pollutants, such as the acid gases sulphur dioxide and the nitrogen oxides.

6.3 Environmental Impact Assessment

Wave energy, like all energy producing technologies, has the potential to produce unacceptable environmental impacts. However, as shown above, prior evaluation (e.g. environmental impact assessment), careful selection of the place and time of deployment, and prior consultation with interested parties should minimize any critical environmental impacts. Such considerations could (in time) form the basis of "best practice" for deploying wave energy schemes. For the present, a checklist of the most important subjects to be considered has been devised by Hideo Kondo (Kondo 1998), which is designed to be completed according to the following categories: (a) may improve; (b) small damage; (c) partial damage; and (d) heavy damage (Table 6.3).

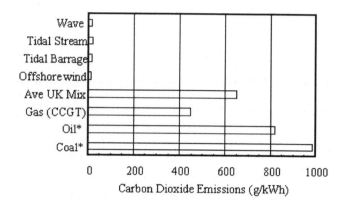

Figure 6.1 Comparison of life-cycle emissions of carbon dioxide. (*– assumes best modern practice.) Source: T.W. Thorpe, Energetech, Australia. (Reprinted by permission.)

| ENVIRONMENT FACTOR | LAND | | | COASTAL | | OCEAN | |
SITE DEVICE	Physical	Biological	Visual	Physical	Biological	Physical	Biological
Land Connected *Tapered Channel* *OWC (Bottom Fixed)*							
Near-shore *OWC (Bottom Fixed)* *Mov. Body in Chamber*							
Offshore *OWC (Floating)* *Flexible Bag* *Moving Body*							

Table 6.3, Checklist for assessing the effect of wave-power installations on coastal and marine environments
Source: H. Kondo, Coastsphere Systems Institute, Japan. (Reprinted by permission.)

6.4 Socioeconomic Considerations

Applications of wave energy

Wave energy has the potential to make a significant contribution to the total national electricity supply of those countries having suitable wave climates along their coasts. For example, several of the European states that have received European Commission funding to promote regional development have large wave energy resources (e.g., Ireland, Scotland and Portugal). According to Hagerman (1995a), wave energy conversion devices must be proven in electricity distribution networks designed to meet local demands on the coast before they can be seriously considered for central station applications aimed at exporting bulk power to the main grid. In the same paper, Hagerman notes that the economic viability of such small power plants can be enhanced by the variety of co-products that can be obtained incidental to the generation of electricity, including production of fresh water for household use, livestock watering, and crop irrigation. In the case of many developing-countries with coastal regions, wave energy conversion has the potential to increase regional development and the standard of living, since it would remove the requirement for electrical generation based on expensive imports of fuels (e.g., diesel). This would, in turn, lead to greater wealth being retained by local communities and would likely increase opportunities for employment.

The wide availability of wave energy resources can foster the development of self-sufficient island communities (and of other isolated communities). For many remote island communities and isolated coastal communities, problems of water and energy shortages have always existed. Few of the more than 100,000 inhabited islands throughout the world have either adequate water supplies or readily-available reliable sources of energy. Many islands rely on imported water or reverse osmosis (RO) desalination systems for potable water, and on fossil fuels for their energy. In many instances, rain occurs for about six months each year, while the remaining six months are at near-drought conditions. Because RO systems require significant power for their operation, the potable water produced by RO systems is expensive. For example, in some island communities, potable water costs up to $1.06 per litre, primarily due to energy costs which exceed $0.45 per kWh.

Population increases and higher living standards in developing countries will result in considerably increased demands for water with consequent pressures on available resources. This trend is reflected by an ever increasing growth in the sales of desalination plants (International Desalination Association 1998; Wangwick 1998). The United Nations Environment Programme (UNEP) sees the struggle to ensure adequate water supplies as being one of the most important issues facing the world in the 21st century. Wave energy is particularly well positioned to play an important part in meeting this challenge in countries with arid or semi-arid coastlines (Fig. 6.2). The desalination of seawater can be accomplished directly (i.e. without the generation of electrical power) by any wave energy device that uses seawater as a working fluid and develops pressures of 5.5-6.9 MPa (800-1000 psi) (Hagerman 1995a). The McCabe Wave Pump (McCormick 2001; McCormick *et al.* 1998) is an example of a wave energy technology designed primarily to produce potable water.

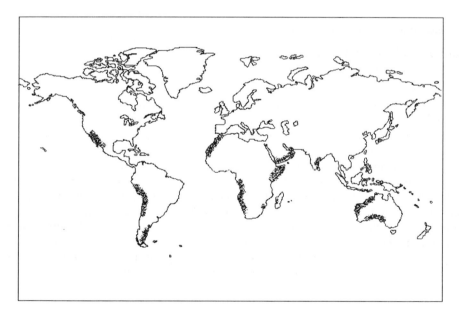

Figure 6.2 Arid, ocean-facing coasts suitable for wave power schemes
Source: T.W. Thorpe, Energetech, Australia. (Reprinted by permission.)

Among other potential applications of wave energy are the following:

- Seawater renewal for closed-pond aquaculture and breakwaters for ports and harbours are two potentially exploitable co-products of wave energy conversion discussed in Hagerman (1995a). Also considered in this paper is a gas utility application of wave power involving the production of hydrogen for use as a winter heating fuel.

- The need for reliable water and energy supplies is of concern to isolated military installations, such as the U.S. Naval Base at Diego Garcia in the Indian Ocean. Although the costs of such supplies are basic considerations, their strategic value is of critical importance. Impedances in energy supply lines can result in compromised military missions. The need for alternative energy capacity for such bases is demonstrated in a 1997 advertisement in the technical press (NAVFACCO 1997).

- In some places in the world, wave-power devices could be used for direct exchange and aeration of stagnant bottom water. Examples are some fjords with shallow sills in Scandinavia, the Baltic proper, and the Black Sea, where less-salty or brackish water overlies relatively heavy, cold sea water. This condition (stratification) inhibits mixing, which results in oxygen deficits and sea-floors devoid of life (Claeson & Sjöström 1988; Czitrom 1997).

- Although not strictly an application, it should be noted that the development of wave energy in the longer term could help to address some of the concerns arising from global warming (International Energy Agency 1994b).

Industrial benefits

Wave power can be exploited using materials that are widely available (primarily concrete and steel). In the long run, basic wave energy infrastructures could be manufactured in almost every region of the world. However, if wave energy is to have a broader future, the utilised concepts must have the capability of mass production and ease of deployment. This criterion applies mainly to near-shore and offshore devices; since most land-based plants will likely have to be designed to meet a variety of physical and social restrictions. In this context, gully-based OWC schemes, and also Tapchan installations, seem to have little prospect for large-scale production because their placement relies on very specific physical conditions. However, some OWC installations (e.g., those based on the Wavegen and Energetech systems) are capable of being incorporated into man-made structures, such as breakwaters and harbour walls, thus extending the potential industrial benefits.

Legal issues

Before wave energy can be economically exploited at open-ocean sites, it would be wise to have international agreements and regulations drawn up to settle the question of who are the owners of wave energy at sea. It should be born in mind that the swells on the shores of a country may result from storms originating thousands of kilometers away. There could be a potential conflict if a nation state were to exploit waves in international waters before the swells reached coastal wave power-plants in the territorial waters of another state. Although such problems seem at present to be very much in the future, it is recommended that appropriate international regulations should be agreed upon well in advance of developments so that they are in place before potential conflicts arise.

Chapter 7

WAVE POWER ACTIVITIES IN THE AMERICAS AND OCEANIA

High wave power levels, of the order of 40 to 60 kW per meter of wave crest length, are experienced along exposed shores of the west coast of North America between latitudes 35 degrees and 60 degrees north (Canada and the US, including Alaska). In South America, levels as high as 70 kW per meter are experienced along the shores of Chile between latitudes 35 degrees and 60 degrees south. Elsewhere in this region, similar levels are experienced along the western exposed shores of New Zealand and Australia, including Tasmania. These high levels, which result from winds blowing predominantly from the west (the Westerlies and Roaring Forties), represent potential for wave power developments. The east coast of the American continent has a more modest wave exposure as compared to the west coast.

Wave energy conversion activities in this large geographic region are currently focussed mainly in the United States, Canada, Mexico, Australia and Fiji. In the United States, research on wave energy conversion began following the oil crisis of 1973 as was the case in several other countries, the main emphasis being on offshore devices such as the DELBUOY. Activities in Mexico and Australia got underway in the 1990s, Mexico with a wave-driven seawater pump, and Australia with an advanced-stage-development electrical generation system based on a parabolic focussing device. While Canada has focussed most of its ocean energy conversion work on tidal power, planning got underway in 2001 for a demonstration wave-power plant on the Pacific coast. Wave power represents an attractive potential energy resource for many island communities in Oceania, and in this regard the South Pacific Applied Geosciences Commission (SOPAC) has carried out an extensive regional wave energy resource assessment in Melanesia and Polynesia, including the waters around Fiji.

7.1 United States

Between 1979 and 1996, the United States Department of Energy (DOE) spent approximately $240 million on its ocean energy program. Of this amount, approximately $1.5 million were devoted to wave-related activities, two-thirds of which was earmarked for the building and testing of a prototype 125 kW McCormick counter-rotating air turbine as part of the International Energy Agency's contribution to the Kaimei project (Hagerman 1996) (refer also to Chapter 8, Section 8.1, Japan).

69

Most wave energy initiatives in the United States have been sponsored by private companies, and have generally been of small hardware scale and duration of ocean testing (Hagerman 1996). An early example is the DELBUOY device. More recently, in 2002, Aqua Energy Group announced plans for a $2.5 million wave power demonstration plant off Wa'atch Point, Makah Bay, on the northwest coast of Washington State (Institute of Energy 2002a). Several utilities and state organizations have funded wave energy studies for their service territories. These include Virginia Power, the North Carolina Alternative Energy Corporation, Pacific Gas and Electric Company, and the State of Hawaii (Hagerman 1996). Various studies have also been undertaken by universities.

Operational wave energy devices[1]

DELBUOY (Freely Heaving Float with Sea-floor Reaction Point[2]). Doug Hicks and Michael Pleass of the University of Delaware developed the offshore DELBUOY system for the desalination of seawater by direct reverse osmosis (RO) (Hicks *et al.* 1988). Since 1982, an evolving series of prototypes has been deployed off the southwest coast of Puerto Rico, where fresh water production was demonstrated at a continuous rate of 950 liters (250 gallons) per day from a single buoy. License to this device is held by CHPT, Inc., a Lewes, USA-based company specializing in the design and fabrication of high-pressure hydraulic and structural components from composite materials. Employing a 2.1 m diameter buoy tethered to a sea-floor anchor by a single-acting hydraulic cylinder (Fig. 7.1), DELBUOY produces a pressure amplification ratio of nearly 2800:1. As a result, small waves can generate a pump pressure of 5.5 MPa (800 psi), which is adequate to develop reverse-osmotic flow. In a typical installation, six buoy/pump moorings supply one RO module, which delivers 5.7 m^3 (1500 gallons) of fresh water per day in wave conditions typical of trade-wind wave climates, i.e., waves 1 m high, having a period of 3-6 seconds.

Advanced stage developments[3]

Several wave energy devices have been produced to an advanced stage of development. One of these, the Wave Energy Converter, developed by Ocean Power Technologies Inc. of New Jersey, is an offshore device consisting of a simple and ingenious mechanical system to drive the generators using mechanical force developed by the wave energy converter. The device is based on the *freely heaving float with sea-floor reaction point[2]* principle. It has efficient power conversion electronics to optimize the generated electricity. Based on modular power units, the generators and electronics are located in watertight compartments of ocean-going buoys. Standard marine-grade power cabling and grid-connection hardware and software are employed. A combination of mooring chains and anchors is used for positioning, while underwater hubs and electrical power cables are used for interconnection and transmission to the shore. The system has been extensively tested at a large scale in the Atlantic, and the first commercial schemes are planned for Australia (refer to Section 7.4, below) and in the Pacific, with a number of other possibilities (Taylor 1999; Thorpe 1999a, 1999b).

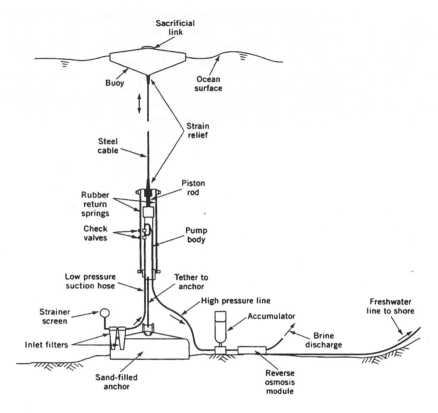

Figure 7.1 DELBUOY device
Source: Yeaple 1989.

Another system produced to an advanced stage of development is the Wave Energy Module (WEM). In the relatively small buoy systems (point absorbers) described above, heave is the predominant energy absorption mode due to the relatively small ratio between float diameter and wavelength. Larger floats not only heave, but also develop angular motion (pitch and/or roll) as they attempt to follow the contours of the sea surface. A leading example of such a wave-contouring float is the Wave Energy Module, which is based on the *contouring float with inertial reaction point*[2] energy conversion process. This offshore system (Fig. 7.2) was invented in the late 1970s by Harry Hopfe, and its development was pursued by U.S. Wave Energy, Inc. of Longmeadow, Massachusetts. A series of tests has been conducted on a 1 kW model in Lake Champlain (Hopfe & Grant 1986). The WEM's operating principle has much in common with that of the Hosepump device (refer to Chapter

9, Section 9.2, Sweden), in that fluid power is generated by the wave-induced motion of a circular float relative to a suspended damper plate. Unlike the Hosepump concept, however, where each buoy is simply an absorber, the WEM buoy contains the entire power conversion system. During their power stroke, the cylinders pump hydraulic fluid to a high-pressure accumulator. When the buoy heaves downward, or when pitch or roll causes one side of the buoy to tilt towards the damper plate, the cylinders are returned to their neutral position by fluid from a low-pressure accumulator and are thus primed for their next power stroke. The high-pressure accumulator discharges fluid to a hydraulic motor that is coupled to an electrical generator. With consistently low oil prices from 1986 to 1998, this project was relatively dormant over this time period. However, with the subsequent rise in oil prices, there has been renewed activity, including: fabrication cost-reduction studies; siting studies with reference to the South Pacific region; and basic performance improvement studies (undertaken by the University of Rhode Island's Ocean Engineering Department).

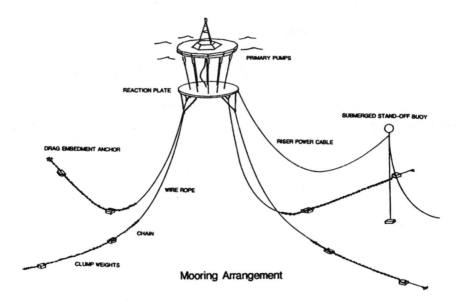

Figure 7.2 Wave Energy Module (WEM)
Source: U.S. Wave Energy Inc., USA. (Reprinted by permission.)

While a circular float such as the WEM follows the sea surface contours by heaving, pitching, and rolling, a long-narrow raft absorbs wave energy primarily from heave and pitch. In the mid-1970s two contouring-raft concepts with hydraulic power take-off were proposed independently and almost simultaneously: one in the United States by Glenn Hagen, and the other in the United Kingdom by Christopher Cockerell, inventor of the Hovercraft. Hagen's Contouring Raft offshore system falls under the *contouring float with sea-floor reaction point* category[2]. Its patent is assigned to Williams, Inc., a Louisiana-based company that formed Sea Energy Corporation to further develop Hagen's contouring raft concept. In the original concepts of Hagen and Cockerell, the latter of which was tested at sea, several floating rafts are hinged together, end-to-end, and oriented to meet incoming waves head on. Passage of a wave causes relative pitching between adjacent rafts, and this motion is converted into fluid power by hydraulic cylinders mounted across the hinges. Sea Energy Corporation's early test results suggested that if a practical means could be found for absorbing energy from the wave-following motions of a single raft, it would absorb energy from both pitch and heave, rather than pitch alone, thereby improving its efficiency. Such a device would require less water-plane area than a multiple-raft unit to capture the same amount of energy in a given sea state. Hull costs would drop by a comparable amount, and the problems associated with hinging several rafts together could be avoided (U.S. Patent 1988). The single-raft concept is described as follows:

- a ship- or barge-shaped hull is connected by a rigid yoke to a reference point fixed in heave, such as a taut moored buoy;
- the yoke is pinned at either end and pierces the raft hull via sealed bearings;
- inside the raft, the hinge pin is fitted with lever arms spaced at regular intervals, each of which is pinned to the rod end of a hydraulic cylinder that is clevis- or trunion-mounted on the raft;
- as the raft pitches or heaves in passing waves, the hinge pin rotates relative to the raft and strokes these cylinders, thereby converting the surface-following motions of the hull into fluid power;
- due to the unsteady nature of raft motion in high and low wave groups, cylinder pumping action is intermittent; and
- large-volume accumulators are incorporated into the hydraulic circuit to smooth these pulses and supply fluid at constant pressure to a hydraulic motor/generator.

Other developments and activities

A private company, Q Corporation, has devised a tandem flap device that consists of two bottom-hinged flaps placed one behind the other in the direction of wave travel. The flaps are mounted in an open-frame platform, which can be installed in deeper water than a caisson. The company has funded a series of scale model tests, and with co-funding from the U.S. Department of Energy, tested a 20 kW prototype in Lake Michigan during the summer and fall of 1987 (Wilke 1989).

The Neptune System is a process in which wave forces distributed over a heaving float's water-plane area are transmitted to a much smaller diameter pump, thereby achieving a pressure increase and

developing sufficient head to run a high-speed water turbine. Originally developed by Wave Power Industries of Arcadia, California, patent rights to this process have since been assumed by Ocean Resources Engineering, Inc., also of Arcadia. In 1987, the Taiwan Power Company undertook a wave energy resource and technology assessment for their service area; as a result, the Neptune System was selected for further investigation (Wu & Liao 1990).

Swedish companies Interproject Service (IPS) and Technocean have been involved in a cooperative project in the United States with Ocean Power Technologies Inc., Princeton, NJ. In January 1998 a small IPS converter with a piezo electric generator was tested off New Jersey. This field test was preceded by testing a 2/5 scale model in a wave tank, which led to some modifications. The piezo electric generator has the advantage of constant efficiency for all rotational speeds.

In 2002, AquaEnergy Group Ltd. of Mercer Island, Washington announced plans for a 1 MW demonstration wave energy power plant to be located at Makah Bay on the northwest coast of the state (Institute of Energy 2002a; AquaEnergy Group 2002). The stated goal is the delivery of power to the local grid by the end of 2003. As of December, 2002, surface measurement devices had been deployed to determine wind and wave intensities over a six-month period. The installation will consist of several moored-buoy wave energy converters (AquaBuOYs), which are based on the IPS Buoy technology (see above).

7.2 Canada

BCHydro, a power company owned by the Province of British Columbia, announced in June 2001 that a 20 MW green energy demonstration project is to be located on Vancouver Island off Canada's west coast (BCHydro 2001). The project is to be made up of approximately 10 MW of wind power, 6-8 MW of micro hydro-power, and 3-4 MW of wave power; with the three components planned to start operating between 2001 and 2003. Subsequently, BCHydro signed memorandums of understanding with two ocean wave energy developers in respect of their respective technologies: Energetech Australia Pty. – oscillating water column system (Institute of Energy 2002a) (refer to Section 7.4, Australia, below, for details of this system) and Ocean Power Delivery, UK – Pelamis energy converter (refer to Chapter 9, Section 9. 7, UK, for details of this system). The two devices are planned to be deployed at sites off the west coast of Vancouver Island.

7.3 Mexico

Activity in Mexico has centered on a wave-driven, resonant seawater pump, developed at the National University of Mexico, which has the potential for various coastal management purposes such as aquaculture, flushing out of contaminated areas, and the recovery of isolated coastal lagoons as fish breeding grounds (Czitrom 1997; Czitrom *et al.* 2000a, 2000b).

Operational wave energy devices[1]

Wave-driven Resonant Seawater Pump (Fixed Oscillating Water Column[2]). This onshore device comprises a resonant duct, a variable volume air compression chamber, and an exhaust duct (Fig. 7.3). The wave-induced pressure signal at the mouth of the resonant duct drives an oscillating flow that spills water into a compression chamber, and exhausts through a duct to the receiving body of water. Maximum efficiency is attained at resonance, when the system's natural frequency of oscillation coincides with the frequency of the driving waves. Resonance is maintained at various wave frequencies by a novel tuning mechanism that is controlled by a programmed electronic device. Flow rates of the system are up to 200 liters/sec., depending on site and wave (and tidal) conditions. The absence of moving parts allows marine organisms to pass undamaged so that the pump can be used for the biological management of coastal water bodies. Development has gone through stages of theory, physical and numerical modeling, and sea trials with a prototype. The trials, which took place on the coast of Oaxaca, Mexico in 1995, involved the successful pumping of fish larvae from the Pacific Ocean, over a sand bar, to a lagoon that is cut off from the ocean. Latest (2001) plans call for the device to be deployed to flush out a contaminated port and to manage fisheries at a coastal lagoon. The cost of installing a pump varies depending on the terrain, wave conditions and other considerations. A typical installation is likely to cost around US$250,000.

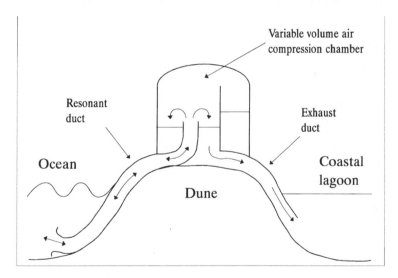

Figure 7.3 Wave driven seawater pump
Source: S.P.R. Czitrom, National University of Mexico. (Reprinted by permission.)

7.4 Australia

In 2001, Energetech Australia Pty Ltd signed a Memorandum of Understanding with Primergy Ltd, a leading Australian renewable energy company, to establish a joint venture company aimed at developing wave energy projects in Australasia and the Asia-Pacific region. The teaming of the two entities is intended to ensure a rapid commercialization of wave energy technology.

Advanced stage developments[3]

Port Kembla Parabolic Focusing Device (Fixed Oscillating Water Column[2]). This Energetech development is a shoreline device suitable for locations where there is fairly deep water right up to the coast, such as on harbour breakwaters and at rocky headlands and cliffs. It utilizes about 40m of coastline. The system has two novel concepts: a turbine suited to the oscillating airflows in the oscillating water column (OWC); and a parabolic-shaped reflector to concentrate the wave resource on the OWC chamber (Thorpe 1999a, 1999b; Energetech 2000; Green 2000). The device, which will generate a maximum of 500kW, is to be installed in Port Kembla, New South Wales in 2003 (Fig. 7.4). Parabolic focussing was first proposed in 1990 and then, in 1992, concepts of an energy extraction chamber and a new design of turbine (see below) were added to form the Energetech Wave Energy System. The concept testing phase of the development, using a scale model, was completed in 1997. At the focus of the parabolic wall, the water is predicted to rise and fall periodically with an amplitude of two to three times that of the incoming waves (all the energy of the incoming plane wave impacting on the parabolic wall converges on this point). This amplified vertical movement is captured by a steel OWC chamber, which connects to the mechanical and electrical plant situated behind the OWC chamber. The installed Denniss-Auld turbine, like the Wells turbine, can cope with oscillatory air flow but operates at lower rotational speed and higher torques in order to improve efficiency. The turbine uses a sensor system with a pressure transducer, which measures the pressure exerted on the ocean floor by each wave as it approaches the capture chamber, or as it enters the chamber. The signal from the transducer is sent to a programmable logic controller (PLC) that adjusts (in real time) the various control parameters of the tubine (e.g., blade pitch) and generator (c.g., speed and torque characteristics of the generator load) in order to maximize the power transfer. The model and concept testing phase of the Energetech development was completed in 1997 at the University of New South Wales Water Research Laboratory in Manly Vale. The turbine was initially tested as a theoretical and numerical concept on computer-generated models at the Department of Aeronautical Engineering at the University of Sydney, and has been constructed at full-scale by Moss Vale SMS.

A second wave power system of interest in Australia is the OPT Wave Power System referred to under United States (Section 7.1), above. As previously noted, it is an offshore system produced by Ocean Power Technologies, USA to an advanced stage of development[3]. Based on the *freely heaving float with sea-floor reaction point concept[2]*, it is in operation at a number of demonstration sites around the world (Taylor 1999). It is based on modular power units resembling large buoys that convert water motion into electrical power for transmission to the grid. The generator units compensate for variations in tide and wave height, and automatically protect against large wave forces. The Australian company involved, Ocean Power Technologies (Australasia) Pty Ltd., is currently (2001) working on a 20kW development project to be established at Portland on the southern coast of Australia in conjunction with Powercor, an Australian utility.

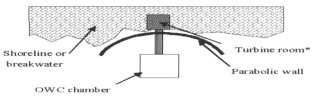

Figure 7.4 Port Kembla parabolic focussing device
Source: Energetech Australia Pty. Ltd. 2000. (Reprinted by permission.)

7.5 Fiji

A UN-sponsored wave energy project to implement a 5kW power plant is underway in Fiji. The United States company Ocean Power Technologies Inc. of New York is involved (refer to Section 7.1 for information on this company). Also with reference to Fiji, the multi-national organization SOPAC, based in Suva, cooperated with the Norwegian company OCEANOR in the mapping of the South Pacific wave climate (Barstow & Falnes 1996) (refer also to Chapter 9, Section 9.1, Norway).

---

1. Category comprises full-scale devices, chiefly prototypes, that are currently operating (or have operated) where the energy output is utilized for the production of electricity or other purpose; also includes full-scale devices at an advanced stage of construction.

2. Refer to Table 3.1, Chapter 3, for details of the wave energy conversion process classification system used in this book; the system is a modified and updated version of that developed by George Hagerman (see Fig. 3.1, Chapter 3; also refer to Hagerman 1995a).

3. Category comprises: (a) devices of various scales, including full-scale, that have been deployed and tested *in situ* for generally short periods but where the energy output has not been utilized for the production of electricity or other purpose (in most cases plans call for the systems to be further developed and deployed as operational wave energy devices); and (b) full-scale devices planned for construction where the energy output will be utilized for the production of electricity or other purpose. Note: devices at an early stage of development are not included.

Chapter 8

WAVE POWER ACTIVITIES IN THE ASIA-PACIFIC REGION

The Pacific-facing coastline of Asia has a more modest wave exposure than that of the west coast of the Americas, e.g. 10 to 15 kW per metre of wave crest length off the coast of Japan. Nevertheless, there is a significant level of wave energy conversion activity, due in part to the large population with many potential customers, including island nations such as Indonesia, Philippines and Sri Lanka.

Japan, a pioneering nation in the wave energy conversion field, has the largest program. Its work dates back to the 1960s with innovative research on navigation buoys, and continues today with several operating plants producing electrical energy. Most recently in Japan, a floating oscillating water column device has been constructed and deployed with the dual purpose of producing electricity and providing shelter. China and India, whose ongoing research programs in the wave energy conversion field started in the 1980s, each have one operational plant. South Korea is involved in several projects, including a combined wave-wind power generation device. Among other countries, Russia has several institutes with research and development programs in the field, Indonesia is involved in research and planning initiatives, and the Philippines has had a survey of its wave energy resource undertaken.

8.1 Japan

There has been substantial research on wave energy conversion in Japan. As far back as 1947, Yoshio Masuda, who still – more than 50 years later – remains active in the field, conducted his first sea trials. He is now associated with the Ryokuseisha Corporation, which markets wave-powered navigation buoys and also pursues development of the "backward bent duct buoy" (BBDB) device. Around 1980, when Masuda was an employee of JAMSTEC (Japan Marine Science and Technology Center), the Kaimei floating platform project was conducted by Japan with participation from International Energy Agency (IEA) members Canada, Ireland, Norway, Sweden, the UK and the USA. JAMSTEC established a major wave energy research and development program in the early 1970s, which continues today, and which has led to the development of several wave energy devices.

Operational wave energy devices[1]

Shoreline Gully, Sanze (Fixed Oscillating Water Column[2]). Shoreline gullies are naturally tapered channels, and an oscillating water column (OWC) at the head of such a gully is exposed to higher wave power densities than those found at the gully's mouth. Systems based on this principle have been built in several countries, primarily for testing pneumatic turbine designs. The Japanese version (Hotta *et al.* 1986) was a 40 kW onshore unit that operated for six months at Sanze on the west coast of Japan before it was taken out of service in 1984.

Offshore Breakwater, Sakata Port (Fixed Oscillating Water Column[2]). Another Japanese onshore fixed OWC system is based on caissons placed side-by-side in a breakwater configuration. Such an OWC has been developed by the Japanese Ministry of Transport under the direction of Yoshimi Goda, and a 60 kW prototype has been installed as part of a new offshore breakwater built at Sakata Port on the west coast of Japan (Fig. 8.1) (Ohno *et al.* 1993). The breakwater consists of a row of caissons on a rubble mound foundation, one of the caissons being built with a "curtain wall" that forms the OWC capture chamber.

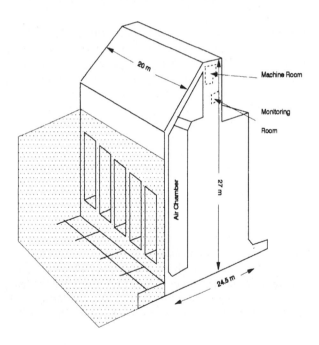

Figure 8.1 Sakata Port OWC caisson
Source: T.W. Thorpe, Energetech, Australia. (Reprinted by permission.)

Kujukuri (Fixed Oscillating Water Column[2]). A full-scale, onshore, wave power generating plant incorporating a constant air-pressure tank system has been built at the Kujukuri, Chiba Prefecture (Fig. 8.2) (Hotta *et al.* 1996; Miyazaki *et al.* 1993). With a maximum output of 30 kW, it is used as a supplementary power source for a flounder farm. A bank of wave energy converters, which convert wave energy to air pressure, are located on the seaward side of a seawall. A constant air-pressure tank equalizes fluctuating air pressures transmitted from the converters, and provides a flow of air at constant pressure to an air turbine. The wave energy converters also serve as wave energy dissipation structures.

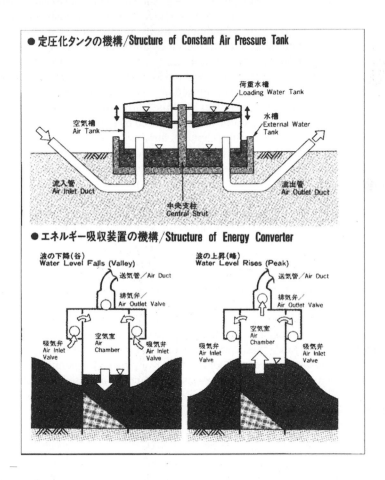

Figure 8.2 Kujukuri OWC
Source: Japan Marine Science and Technology Center. (Reprinted by permission.)

81

Haramachi (Fixed Oscillating Water Column²). A prototype onshore 130 kW OWC was deployed at Haramachi in 1996 (Hotta *et al.* 1996). The installation uses rectifying valves to control the flow of air to and from the turbine in order to produce a steady power output.

Caisson-based "Pendulor", Muroran Point (Pivoting Flaps²). The Muroran Institute of Technology, Hokkaido, has developed a caisson-based, pivoting flap device named the "Pendulor" system (Watabe & Kondo 1989). The device utilizes a high-pressure oil power take-off system. In April 1983, a 5 kW (hydraulic motor rating) onshore prototype was installed at Muroran Port on the south coast of Hokkaido. The prototype caisson was sited in front of an existing seawall in a water depth ranging from 2.5 m at low tide to 4 m at high tide. Two capture chambers were built into the caisson, but only one was fitted with a Pendulor. Twenty months after its installation, the Pendulor was bent during a severe storm; as a result, the shock absorbers for the end-stops, which prevent over-stroking of the cylinder, had to be redesigned. A new Pendulor was installed in November 1985, which survived several severe storms without damage. In addition, a small Pendulor system for the generation of electric power was deployed in 1981. Rated at 20 kW, this unit was used to heat the public bath of a fishing cooperative at Mashike Harbor on Hokkaido's west coast. However, its Pendulor was also damaged by a storm. It was replaced by a shorter Pendulor in 1983, which left a considerable gap at the bottom of the capture chamber, and while this has prevented further damage, it has also lowered the system's conversion efficiency. Nevertheless, the plant continues to operate. More recently, and on the basis of the above experience, a new design has been developed for a larger 300 kW pendular device (Fig. 8.3).

Caisson-based, Wakasa Bay (Pivoting Flaps²). The Kansai Electric Company has developed an alternative caisson-based, pivoting flap device in which the flap is hinged at the bottom rather than top, and a 1 kW onshore test unit has been installed on Wakasa Bay, northwest of Kyoto (Miyazaki 1991).

Navigation Buoys (Freely Floating Oscillating Water Column²). This wave energy conversion process was the first to achieve widespread commercial application. As a result of research by Yoshio Masuda in Japan, a battery-charging generator was developed for navigation buoys in the near-shore to offshore zone, driven by the oscillating water column of a central pipe in the buoy's hull (Fig. 8.4). Since Masuda's developmental work in 1964-65, approximately 700 such generators (rated at 60 watts) have been sold by the Ryokuseisha Corporation for use in Japan, and another 500 have been exported to other countries. Similar units are also being produced by Munster Simms Engineering, Northern Ireland, UK, and the Guangzhou Institute of Energy Conversion in the People's Republic of China (see Section 8.3, China).

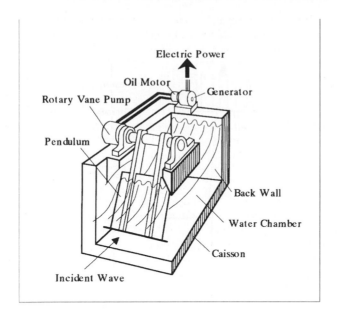

Figure 8.3 300 kW Pendular device
Source: Japan Marine Science and Technology Center. (reprinted by permission.)

Kaimei Floating Platform (Freely Floating Oscillating Water Column^2). Yoshio Masuda also pursued the development of OWC technology for larger floating power plants, beginning with his work at the Japan Marine Science and Technology Center (JAMSTEC) on the test ship *Kaimei*, and more recently at the Ryokuseisha Corporation on the Backward Bent Duct Buoy. JAMSTEC began its wave energy research and development program in 1974 and, following two years of laboratory testing, had the *Kaimei* built as a prototype floating platform for testing relatively large (up to 125 kW) pneumatic turbine-generators in the near-shore to offshore zone. The *Kaimei* has a length of 80 m, a beam of 12 m, and a design draft of 2.15 m. Thirteen open-bottom capture chambers are built into its hull, each having a waterplane area of 42 to 50 m^2. The *Kaimei* was deployed twice in 40 m water depth off the west coast of Japan near the port of Yura. During its first deployment, from August 1978 to March 1980, eight turbines were tested aboard the vessel, all using various arrangements of non-return valves to rectify the air flow. For four months during the winter of 1978-79, wave-generated power from one of the impulse turbines was supplied to the mainland grid. During its second deployment, from July 1985 to July 1986, five turbines were tested, including three impulse turbines (all with rectifying valves), a tandem Wells turbine, and a McCormick counter-rotating turbine (Hotta *et al.* 1988).

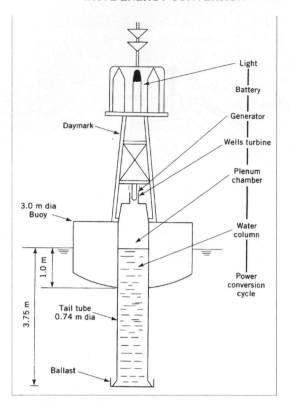

Figure 8.4 Floating OWC navigation buoy
Source: Japan Marine Science and Technology Center. (Reprinted by permission.)

Mighty Whale (Fixed Floating Oscillating Water Column[2]). Floating OWC devices have been developed in Japan that are stable in pitch and heave, so that their performance is more like that of a caisson-based OWC. Unlike caissons, however, such floating platforms can be moored in deeper water, taking advantage of the greater wave energy resource. The first of these stable floating OWC devices, the Mighty Whale, was developed at JAMSTEC under the leadership of Takeaki Miyazaki for deployment in the near-shore to offshore zone (Figs. 8.5 & 8.6) (Hotta *et al.* 1996). It is a development of the Backward Bent Duct Buoy (BBDB) (see "Other developments and activities", below, for details). OWC capture chambers line the front of the device, with buoyancy chambers behind these. A flat ramp slopes down and back from the capture chambers into the water, damping the pitching motion of the device. Wave tank tests have shown that the device has a maximum capture ratio of about 70%, together with a wide capture bandwidth and small mooring

84

forces (Miyazaki *et al.* 1993). In addition, the efficient wave absorption of the device gives rise to relatively calm waters behind it, which provide good locations for fish farming, marine sports, etc. This combination of effects establishes the Mighty Whale as a promising candidate for the dual requirements of generating power and supporting coastal mariculture. The Mighty Whale is a joint development of JAMSTEC and Ishikawajima-Harima Heavy Industries Co. Ltd. (IHI). The first prototype was constructed at IHI's Aioi Works, and in 1998 was delivered to JAMSTEC and moored at a test site off Gokasho Bay, Mie Prefecture. The device is 50 m long and 30 m wide, which makes it the world's largest wave-power device. It carries one 10 kW, one 50 kW and two 30 kW generators.

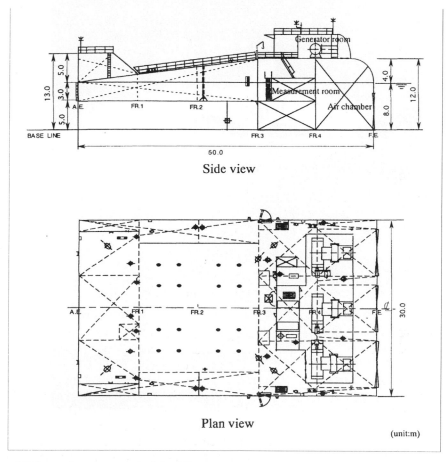

Figure 8.5 General arrangement of the Mighty Whale
Source: Japan Marine Science and Technology Center. (Reprinted by permission.)

Figure 8.6 Operation of the Mighty Whale (diagrammatic representation)
Source: Japan Marine Science and Technology Center. (Reprinted by permission)

Advanced stage developments[3]

An offshore hydraulic conversion system has been developed to an advanced stage by the Japan Institute for Shipbuilding Advancement (Yazaki *et al*. 1986). In 1984-85, prototype tests were conducted aboard the specially-built jack-up rig, *Kaiyo*, off Iriomote Island, southwest of Okinawa. Based on the *contouring float with sea-floor reaction point*[2] wave energy conversion process, two floats were located in the open bays of *Kaiyo* and were linked to the rig in such a way that heave, pitch, or surge of the floats resulted in movement of pistons in the hydraulic cylinders. Unlike the US Sea Energy Corporation's conversion system (refer to Chapter 7, Section 7.1, United States), where fluid is pumped to accumulators that are simultaneously charging and discharging, the accumulators on *Kaiyo* were arranged in pairs. One member of each pair was charged over a period of time (approximately 10 minutes), while the other member discharged fluid to a hydraulic motor/generator. When the charging accumulator reached operating pressure as a result of the pumping action of the floats, the accumulators were switched.

Other developments and activities

A major research and development initiative is devoted to the improvement of the floating OWC process through the development of the Backward Bent Duct Buoy (BBDB). This device absorbs wave energy from both the heave and pitch of a ship-shaped hull, although pitch appears to be the most important absorption mode. For a given capture chamber area and wave height, the BBDB absorbs three times the power of a center-pipe navigation buoy, and ten times that of the *Kaimei* (Masuda *et al*. 1987). Yoshio Masuda is a key researcher in the field, his most recent work having been carried out in cooperation with researchers at the Guangzhou Institute of Energy Conversion, People's Republic of China. In this program, wave tank tests indicated that the addition of a half-cylindrical body just aft of the BBDB's riser section greatly improves the performance of the device. Specifically it leads to a higher and broader peak in energy absorption efficiency, and a shifting of this peak to longer wave periods (so that a shorter, less costly hull can be used in a given wave climate). As noted previously, the Mighty Whale initiative is a development of the BBDB.

Among other Japanese research and development activities, the Taisei Corporation has conducted laboratory and mathematical modelling of an OWC tension leg platform (Tanaka *et al*. 1993), which is stable in both pitch and heave. This makes it an ideal candidate device for powering floating airports and other offshore platforms requiring good stability. In related work, a float-in-caisson pump has been developed by Kajima Corporation (Shiki & Iwase 1990), but has not been tested at sea.

8.2 South Korea

Advanced stage developments[3]

The Korea Research Institute of Ships and Ocean Engineering (KRISO), Daejon, has developed to an advanced stage a *fixed floating oscillating water column device*[2], which comprises a 13 m cylindrical structure with four mooring units. With a rated power output of 40 kW, the device uses a tandem Wells-type turbine system for energy pick up. Hong (2001) reports that the device was launched for sea tests in mid-2001.

Other developments and activities

A paper presented at the 1998 PACON Congress states that the Korean firm Baek Jae Engineering has developed an initial design concept for a prototype wind-wave energy scheme (Fig. 8.7) (Cho & Shim 1999).

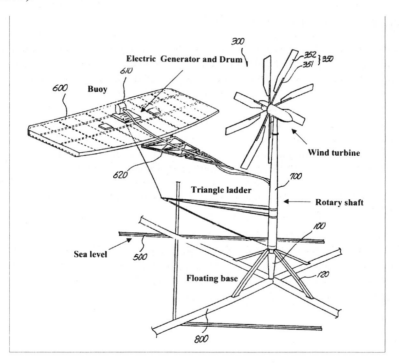

Figure 8.7 Outline design of combined wave-wind power generation device
Source: Cho, Kyu-Bock & Shim, Hyun-Jin 1998.

Featuring a large floating base of pipes in the form of a lattice, the pipes contain pressurized air resulting in a buoyant structure that is moored via an anchoring system. The lattice serves to tie a large number of buoys together, holding them at the required separation distance and serving as a reaction frame against which they oscillate up and down. Electricity is generated as a result of this oscillation, and also from wind turbines added to the structure. It is claimed that the device could be economically competitive with a range of electricity generation technologies if deployed in energetic wave climates. For example, electricity costs for a scheme comprising 956 buoys in UK waters are expected to be in the region of 6-9 c/kWh. Baek Jae Engineering is continuing its R&D with the ultimate object of the construction and deployment of a prototype.

8.3 China

Early work focussed on the use of wave energy for the propulsion of vessels and for powering navigation buoys. Activity in the wave energy conversion field has increased in recent years, with research currently being carried out at more than ten universities and other research institutions. Most fundamental research is supported by the Nature Science Fund of China and the Chinese Academy of Sciences. In the applied field, most projects are the responsibility of the Guangzhou Institute of Energy Conversion (Guangzhou Institute of Energy Conversion 2000). Much of the work of this institute is concerned with oscillating water column (OWC) converters in fixed or floating structures, and is supported by the State Science and Technology Committee on Wave Power Utilization.

Operational wave energy devices[1]

Dawanshan Island (Fixed Oscillating Water Column[2]). An experimental 3 kW shoreline oscillating water column (OWC) device was installed on Dawanshan Island in the Pearl River estuary in 1990 (Yu 1995). Following initial good performance, it was upgraded to operate as a power station with a 20 kW turbine.

Navigation Buoys (Freely Floating Oscillating Water Column[2]): As noted in Section 8.1, Japan, the Guangzhou Institute of Energy Conversion manufactures the Japanese-developed 60W battery-charging generator for navigation buoys deployed in the near-shore to offshore zone. Some 650 units have been produced in the past 13 years. Most are in use along the Chinese coast, with a few being exported to Japan (Yu 1995).

Advanced stage developments[3]

An onshore OWC scheme at an advanced stage of development, and based on a design by the Guangzhou Institute of Energy Conversion, is proposed to be installed at a site near the city of Shanwei. Originally the scheme was to have been installed at Nan-Ao Island, but plans subsequently changed. Consideration is being given to an OWC device with a total capacity of 100kW (Yu & You 1995).

Other developments and activities

Research projects supported by the Nature Science Fund of China and the Chinese Academy of Sciences have addressed the following topics:

*   turbine in oscillating air flow;

*   wave load safety design of wave power devices;

*   time domain modelling and control;

*   non-linear hydrodynamic simulation; and

*   information systems on wave energy resources.

Supported by the State Science and Technology Committee on Wave Power Utilization is an on-shore project by the Tianjing Institute of Ocean Technology based on a fixed pendulum system, which was installed on Daguan Island, Shandong Province in 2000. Other work includes research on a backward bent tube duct buoy (BBDB) wave power converter (Liang & Wang 1996) (refer also to Section 8.1, Japan).

8.4 Russia

Several papers on Russian activities in the wave energy field were presented at the 1999 PACON Congress in Moscow. Among the institutes involved in research and development activities are: Moscow Power Engineering Institute; Moscow State Civil Engineering University; and the P.P. Shirshov Institute of Oceanology, Moscow. One paper outlined work carried out at the Moscow State Civil Engineering University on inertial type wave pumps.

8.5 India

Wave energy research in India started in 1982. It was initially organized under the Wave Energy Group of the Indian Institute of Technology (IIT), Madras. However, in 1995 the new National Institute of Ocean Technology (NIOT) took over responsibility for wave energy activities. After about six years of laboratory research, IIT Madras coordinated the design and construction of a 150 kW wave energy converter of the oscillating water column (OWC) type, similar to the Kværner OWC device built in Norway in 1985 (refer to Chapter 9, Section 9.1, Norway). It is located at Vizhinjam Fisheries Harbour near Trivandrum.

Operational wave energy devices[1]

Vizhinjam Harbour (Fixed Oscillating Water Column[2]). A 150 kW prototype, multi-resonant, OWC with protruding walls was built onto the breakwater of the Vizhinjam Fisheries Harbour, near Trivandrum in 1991 (Fig. 8.8) (Ravindran *et al.* 1997). This onshore scheme, which comprises a Wells turbine coupled to a 150 kW squirrel cage induction generator, functioned well, producing data that are being used to design and build an improved demonstration scheme at the same site having the following features:

- Slip-ring, variable-speed induction generator in place of the original squirrel cage induction generator for improved performance under fluctuating load.

- Two power modules (each having a capacity of 55 kW) in place of the original single module for the purpose of addressing large (15 kW) windage losses that have to be supplied from the grid under low wave energy conditions. In the new scheme only one of the two power modules will run under low power conditions.

- Replacement of the original fixed-chord blade turbine by one with a tapered chord for improved efficiency.

A modified scheme based on the above features, but with only one of the new power modules, was installed in 1996 for subsequent operation and testing (National Institute of Ocean Technology 2000).

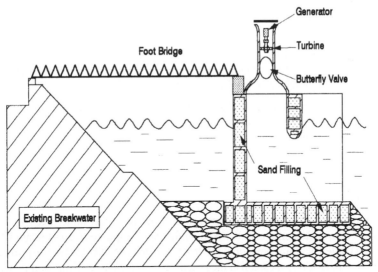

Figure 8.8 Cross-section of Indian breakwater device
Source: T.W. Thorpe, Energetech, Australia. (Reprinted by permission.)

Other developments and activities

Research and development activities pertaining in part to the Vizhinjam OWC device have involved the testing of three different concepts for the wave energy power module:

(a) twin Wells turbine coupled to a 150 kW squirrel cage induction generator (Ardhendu *et al.* 1997);

(b) twin horizontal-axis Wells turbine coupled to a slip-ring induction generator; and

(c) impulse turbine with linked guide vanes coupled to a 55 kW slip-ring induction generator (Santhakumar *et al.* 1998).

NIOT has also completed model studies on the hydrodynamic performance of a floating OWC of the backward bent duct buoy (BBDB) type. Based on the earlier work of Yoshio Masuda of Japan and research in China, a prototype dimension best suited to the wave climate of the Indian coast was determined; as a result, a 1:13 scale model was constructed and tested (National Institute of Ocean Technology 2000).

8.6 Indonesia

On the basis of operating experience at the Toftestallen, Norway, tapered channel plant, plans were finalized for a commercial 1.1 MW plant at Java, Indonesia (Tjugen 1996). However, preparatory construction work at the site was discontinued after the Indonesian financial crisis of 1997 (Anderssen 1999).

As of 2001, research on an oscillating water column device involving model tests in a wave tank was underway in a joint project between Gajah Mada University, Jogjakarta and the Sepuluh Nopember Institute of Technology, Surabaja (ITS) (Suroso 2001).

Also in Indonesia, the government has designated the Baron coastal area of Jogjakarta as an alternative energy park for various forms of renewable energy, including wind power and wave power. One of the research projects involves a combined wave/wind-powered device. (Suroso 2001).

8.7 Philippines

An investigation of the wave energy resource of this country was undertaken in 1996. It was executed by the Norwegian company OCEANOR for the Philippines Department of Energy, with funding from the United Nations Development Program (OCEANOR 2002). The primary objectives of the project were to assess and establish the potential of wave energy for power generation in this country, and to determine the technological viability, economic feasibility and environmental desirability of the utilization of this energy source.

---

1. Category comprises full-scale devices, chiefly prototypes, that are currently operating (or have operated) where the energy output is utilized for the production of electricity or other purpose; also includes full-scale devices at an advanced stage of construction.

2. Refer to Table 3.1, Chapter 3, for details of the wave energy conversion process classification system used in this book; the system is a modified and updated version of that developed by George Hagerman (see Fig. 3.1, Chapter 3; also refer to Hagerman 1995a).

3. Category comprises: (a) devices of various scales, including full-scale, that have been deployed and tested *in situ* for generally short periods but where the energy output has not been utilized for the production of electricity or other purpose (in most cases plans call for the systems to be further developed and deployed as operational wave energy devices); and (b) full-scale devices planned for construction where the energy output will be utilized for the production of electricity or other purpose. Note: devices at an early stage of development are not included.

Chapter 9

WAVE POWER ACTIVITIES IN NORTHERN EUROPE

Exposed to North Atlantic Westerlies, this geographic region exhibits wave power levels of 60 to 70 kW per meter of wave crest length off the west and northwest coasts of Ireland and Scotland, and slightly less off the west coast of Norway. This favourable wave climate has resulted in national wave energy conversion programs in Norway, Sweden, Denmark, the United Kingdom and Ireland. In addition to national funding, in 1992 the European Union launched a research and development program on wave energy conversion.

The Scandinavian countries of Norway and Sweden initiated substantial programs in the late 1970s, although there had been some previous work in the field. In the case of Denmark, a formal program was initiated in the 1990s. An example of the success of the Scandinavian programs is the Norwegian tapered channel prototype plant at Toftestallen, which started feeding electricity into the grid in 1985. Limited research and development effort commenced in Germany in the 1970s and continues to the present. In Belgium, investigations of a floating point absorber system were conducted several years ago. Activity in the Netherlands has centered on a commercial development, the Archimedes Wave Swing device, and there are plans for a prototype to be tested in the ocean in 2003. The United Kingdom has a large government-sponsored R & D program. What is claimed to be the world's first commercial wave power plant, LIMPET (500 kW), began operation on the Scottish island of Islay in 2000; the project has a 15 year contract to supply electricity to the grid. In Ireland, a system called the McCabe Wave Pump has been designed specially for remote island communities to produce drinking water by reverse osmosis and/or electrical power.

9.1 Norway

Operational wave energy devices[1]

In 1978, the Norwegian Royal Ministry of Petroleum and Energy started to fund a substantial R&D program on wave energy conversion. It involved two main projects, both of which were based on (then) recent theoretical developments by Norwegian scientists (see following sections for details of these projects). One project, led by the Central Institute for Industrial Research (SI), involved wave focusing by means of submerged structures, acting in analogy with optical lenses (Mehlum 1982). Some years later, the Tapchan device was proposed for conversion of the focused wave energy. A new company, NORWAVE, took part in the resulting project. The Norwegian Institute of Technology (NTH) and the company Kvaerner Brug AS co-operated on the other project, the development of a phase-controlled power buoy (point absorber), a device originally proposed by Kjell Budal. From 1980, Kvaerner Brug's efforts in wave energy were concentrated on developing a bottom-standing OWC (Ambli *et al.* 1977, 1982). In 1981, as part of the overall R&D program, three different wave-energy plants, each of 200 MW capacity that were envisaged to be built on the west side of the island of Bremanger (61.8 °N, 4.7 °E), were assessed (Olje-og energidepartementet 1982). In 1985, however, a much less ambitious plan was realized on the west side of the island of Toftøy (60.5 °N, 4.9 °E).

Governmental funding for wave energy was at a maximum in 1980 (NOK 16 million), and decreased substantially in the following years.

Operational wave energy devices[1]

There are two operational wave energy devices in Norway, a multi-resonant, oscillating water column system and a tapered channel system, both located at Toftestallen, Toftøy (40 km NW from Bergen) on Norway's west coast.

Multiresonant OWC, Toftestallen (Fixed Oscillating Water Column[2]). Kvaerner-Brug A.S., a large Norwegian hydro-power company, has developed a multi-resonant OWC system to be based on land or on free-standing caissons. Kvaerner-Brug's early work on the oscillating water column concentrated on developing a means to adjust the natural frequency of the water column in order to tune the device as the dominant wave period changes from sea state to sea state. In 1980, the approach was shifted to the design of an absorbing structure that would resonate at several frequencies within the range of wave periods expected at a potential plant site (Malmo & Reitan 1986). It was thought to be more cost-effective for a device to have several fixed resonant frequencies rather than a single, continuously variable one. The corresponding structure designed by Kvaerner-Brug consists of a capture chamber and a "harbour", formed by extending the side walls of the chamber in the seaward direction (Fig. 9.1). A 500 kW demonstration plant based on this concept was built at Toftestallen, alongside Norwave's Tapered Channel plant (Fig. 9.2) (Bønke & Ambli 1987). The island's cliff wall forms the resonant harbour. The plant operated for four years before being partly destroyed by a severe winter storm. Specifically, the bolts connecting the steel structure to the concrete structure suffered fatigue fracture. After the storm, the turbine and the generator were rescued from 80 m water depth, but the plant has not yet (2003) been repaired.

Tapered Channel, Toftestallen (Reservoir Filled by Direct Wave Action[2]). Invented by a group headed by Even Mehlum at the Centre for Industrial Research, Oslo, the tapered channel system consists of a collector, an energy converter, a reservoir, and a power house. A 350 kW tapered channel power plant commenced operation in 1985 at Toftestallen, on Norway's west coast.

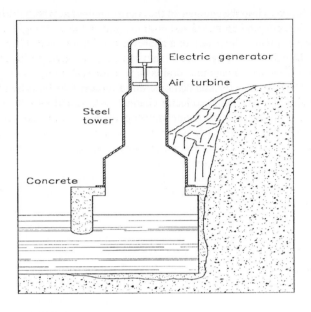

Top: Fig. 9.1 Kvaerner OWC (sectional drawing); Bottom: Fig. 9.2 Kvaerner OWC (Toftestallen demonstration plant)
Source: Fig. 9.1: Falnes 1993; Fig. 9.2: J.Falnes, NTNU, Norway. (Both figures reprinted by permission.)

The collector funnels waves into the entrance of the energy converter, which is a horizontal, vertical-walled channel, having a depth of 6 to 7 m and built in concrete up to a height 2 to 3 m above mean sea level. The channel's width decreases in a shoreward direction, and its end is sealed off. As waves travel along the ever-narrowing channel, they increase in height, spilling water over its sides and into the reservoir (Fig. 9.3). Water then drains back to the sea through a low-head (e.g. Kaplan) turbine/generator. The reservoir for this demonstration project was built by damming two small inlets to the island's interior bay, and a collector channel was blasted into the rock at the head of a natural gully. The reservoir of a tapered channel power plant does not provide long-term storage, but smooths the input from one high-energy wave group to the next. For example, the reservoir at Toftestallen is reported to have an area of 8,500 m^2, while the turbine is designed for a flow rate of 14 to 16 m^3/sec and an operating head of 3 m (Norwegian Royal Ministry of Petroleum & Energy 1987). Should wave energy levels fall so low that waves no longer overtop the channel walls, the reservoir would have its water level lowered by 0.5 m in about 5 minutes. It should be noted, however, that the plant is designed to start automatically whenever sufficient head becomes available again. The tapered channel configuration of the converter enables the device to effectively absorb energy from a large range of wave heights. In Stephen Salter's words (Salter 1989): "Large waves overtop early and deliver a large volume of water. Small waves must travel further along the channel before they get high enough to reach the top of the wall but nevertheless nearly all waves deliver something. This marks the difference between Tapchan and earlier overtopping schemes with walls parallel to the beach." All rights to the Tapered Channel system are held by Norwave A.S., an Oslo-based company. The 350 kW Toftestallen prototype survived several extreme storms, including one that severely damaged the nearby multi-resonant oscillating water column device (see above). However, in 1991 the plant was accidentally damaged in an attempt to improve the shape of its channel and has since not been in operation due to a lack of funding for repairs. On the basis of operating experience at Toftestallen, plans were finalized, through the company INDONOR, for a commercial 1.1 MW plant in Java, Indonesia (Tjugen 1996). However, preparatory construction work on the site was discontinued after the financial crisis in Indonesia during the fall of 1997.

Advanced stage developments[3]

Kvaerner Brug AS (KB) and the Norwegian Institute of Technology (NTH) co-operated on the development of a phase-controlled power buoy (point absorber) with hydraulic power take-off, at first with financial support from the Royal Norwegian Council for Scientific and Industrial Research (NTNF) and then, after 1977, directly from the Royal Ministry of Petroleum and Energy. This wave-energy converter is of the *freely heaving float with sea-floor reaction point*[2] type. In approximately 1981, the proposal involved a power buoy (diameter 10 m), sliding in heave oscillation along a strut connected to an anchored universal joint on the sea floor (depth 40 m). Pneumatic, rather than hydraulic, power take-off was envisaged, and an asynchronous three-phase generator rated at 500 kVA (Budal et al. 1981; 1982). Detailed design work was made, and technical and economic assessments were carried out. Moreover, a model (scale 1:10) was sea tested in the Trondheim Fjord during 1981-1983. The work was discontinued in 1983 because further funding was not made available.

More recently, work on another device, the ConWEC (Controlled Wave Energy Converter) has developed to an advanced stage. ConWEC is a type of wave energy converter that utilizes an oscillating water column and a float in combination. It is based on the *heaving float in bottom-mounted or moored floating caisson[1]* wave energy conversion system, and is intended for deployment in the near-shore to offshore zone. The float is coupled to a pump, which forces water through a turbine (Fig. 9.4). The development project is a collaboration between the Department of Physics, Norwegian University of Science and Technology (NTNU) (Department of Physics, Norwegian University of Science and Technology 2000) and ConWEC AS; it is being supported financially by the Norwegian Research Council (Lillebekken *et al.* 2000). The project has so far included the testing of small laboratory models in a narrow wave channel, testing of larger models in the sea, as well as mathematical simulation and substantial design work. The next stage is the testing of a larger unit on which design work has started. The power capacity of this unit will be approximately 10 kW and will include a turbine. In later stages, larger units may be developed with a capacity of several hundred kilowatts per single unit.

Figure 9.3 Tapered channel device (Tapchan)
Source: Mehlum 1986.

Other developments and activities

After the years 1978-83, with substantial governmental support for developing phase-controlled point absorbers (Falnes 1993), research on optimum oscillation for maximizing the converted wave energy was continued on a more modest scale at NTH (from 1996 at the Norwegian University of Science & Technology NTNU) in Trondheim. A model test and evaluation was undertaken for an optimally operated twin OWC (Falnes 1993), and also a multiple column OWC where several columns with different resonance frequencies spill water into a collecting tube driving one conventional, shared water turbine (Marton 1991). In 1978, a version of a phase-controlled buoy with hydraulic power take-off was tested in the ship model tank at Trondheim (Various authors 1993). This work was continued during the 1990s by Eidsmoen (1996, 1998) undertaking simulation studies. More recently, the Norwegian inventor Tveter (2001) has carried out sea tests off Norway and Denmark on a buoy with hydraulic machinery, the so-called wave pump.

A method was proposed to use wave energy for propulsion of vessels by means of oscillating foils (Jakobsen 1981), and was investigated theoretically (Grue & Palm 1986, 1988) as well as experimentally. In the early 1980s, Einar Jakobsen, Wave Control Co., Norway, demonstrated the effectiveness of a wave propulsion device, a foil propeller, on a 7.5 m yacht hull. This vessel reached a speed of six knots under wave power alone. Tests were then undertaken on the 20.4 m, 180 tonnes, fishing vessel Kystfangst. These demonstrated that in a sea state of 3 m wave height, with horizontal bow foils comprising a total area of 3 m², the foils produced a propulsive force corresponding to 15-20% of the vessel's total resistance at vessel speeds of 4-6 knots. Reduced pitching motion was experienced in head seas and reduced rolling in following seas. (See also Appendix 4.)

Figure 9.4 Principle of the ConWEC device
Source: ConWEC AS, Norway

[KEY Incoming waves (A) hit OWC structure (B); float (C), connected to piston inside pump (E), is put into oscillation; hydraulic fluid (sea water) is pumped from lower level (F) to higher reservoir (G); return of fluid utilized for running turbine (H); latching phase control obtained by clamping two rails (D) during oscillation cycle.]

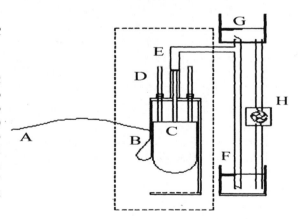

The company OCEANOR has conducted mapping of the wave climate in various ocean regions of the world. Some of the results are given in Chapter 2. The results shown for the South Pacific were obtained in cooperation with the multinational organization SOPAC, based in Suva, Fiji (Barstow & Falnes 1996).

9.2 Sweden

At Chalmers University of Technology, Gothenburg, a wave energy research group was founded in 1977 (Water Environment Transport, Chalmers University of Tecnology 2000). Also, the companies Götaverken Energy, Technocean (TO), and Interproject Service (IPS) (Interproject Services AB 2000) have participated in research and development. Floating offshore converters have been the main subject of study, and small-scale tests were performed during the early 1980s in a lake, as well as tests of several prototypes in the sea near Gothenburg. One of the tested converters is the Hose-pump device (see below), which is expected to be more durable and require less maintenance than an ordinary piston pump device; an alternative design is the so-called IPS buoy.

Advanced stage developments[3]

The Floating Wave-power Vessel (FWPV) is one of three devices developed to an advanced stage. Invented by G. Lagström, the device was originally developed to provide a head of sea water, which was then passed through a filtration system to capture valuable elements from the sea (e.g. gold). It was only in the 1990s that the wave energy capturing potential of the scheme was investigated. The system consists of a floating platform that is single-point moored to always face the prevailing direction of the waves. Based on the *reservoir filled by wave surge[2]* wave energy conversion system, it is planned for deployment in the near-shore to offshore zone. The incoming wavefront encounters a sloping ramp on the platform, and the wave crests spill into collecting basins behind the ramp. The basins are situated above the mean sea-level, and the water flows down through a number of low-head turbines that run electrical generators. Sea Power AB of Gothenburg sea-tested a 110kW pilot FWPV off the Swedish west coast in 1991. It was announced in 1999 that a 1.5MW version of the FWPV is to be constructed in the Shetland Islands, UK (refer to Section 9.7, below).

The Swedish Hose-pump device has been under development since 1980 by Swedyard Corp. (now Celsius Industries), Götaverken Energy, and Technocean (Fig. 9.5) (Sjöström 1994). Now at an advanced stage, the commercial interest in this offshore system has limited the amount of information that is openly available. It is based on the *freely heaving float with inertial reaction point[2]* wave energy conversion process. The hose-pump itself is a specially reinforced elastomeric hose, whose internal volume decreases as it stretches. In the Hose-pump device, the alternative stretching and relaxing is caused by the movement between a float and a damper plate attached to either end of a section of hose-pump. By using one-way valves, sea water is sucked into the hose and then expelled as pressurized water (1-4 MPa) into a collecting line. Output from several Hose-pump devices is manifolded and fed to a hydraulic accumulator, where electricity is generated using a

101

Pelton turbine connected to a generator (this equipment could also be mounted on an offshore platform or be submerged). Laboratory testing of hose-pumps was followed by the installation of a single, small-scale model in lake Lygnern, south of Gothenburg. Later, a larger system, comprising five modules connected to a single turbine and generator, was installed in Lake Lygnern. During 1983 & 1984, a plant consisting three modules and a generator was deployed in the open sea at Vinga off Gothenburg for almost 12 months. One problem was that the elongation of the hose-pump sometimes became too much to survive a long service life. Costs have been derived for a 64 MW station with 360 modules off the Norwegian coast.

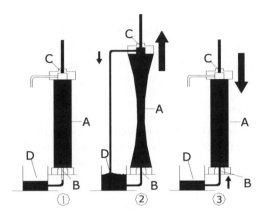

Figure 9.5 Principle of the Hose-pump device
Source: Bengt-Olov Sjöström, Chalmers Univ. of Technology, Sweden. (Reprinted by permission.)

[KEY (1) Hose A is filled with water; (2) Hose has been stretched whereby water is pumped into the collecting line via the check valve C; (3) Hose has resumed its original length & water is being drawn into the hose through check valve B.]

Another development that has reached an advanced stage is the IPS buoy of Interproject Service (Fredrikson 1992). This offshore device is based on the *freely heaving float with inertial reaction point*[2] wave energy conversion process. As the inertial reaction point, the IPS buoy uses a piston contained in an open-ended pipe attached to the buoy hull (Fig. 9.6). A buoy of this type was tested in the sea at Vinga during the summer-autumn period in 1980 & 81. It had a diameter of 3 m, the pipe being 1 m in diameter and 20 m long. The power take-off system was planned as an internal oil-hydraulic system connected to a generator via a turbine or hydraulic motor. Development of the buoy has led to a patented device involving the exchange of the internal oil-hydraulic system for environmentally safer systems. One development is to use the Hose-pump in the IPS buoy instead of the oil-hydraulic cylinder, thus avoiding on the one hand environmental problems with the oil, and on the other, the elongation problem of the hose-pump. A new company (Eurowave Energy AS) is being formed to promote commercial schemes based on the IPS buoy, exchanging the oil-hydraulic cylinder for the hose-pump. This development has grown from the planning of a plant for electricity generation and desalination at the Greek island of Amorgos in the Aegean Sea, a project that has been the subject of several proposals (see Ch.10, Section 10.4, Greece). Deployments are also planned off Australia and in the South Pacific. Full-scale trials with the IPS buoy offshore wave energy converter and the Hose pump system were performed during the period 1980-1986 in the open sea at Vinga, west of Gothenburg, by Interproject Service and Technocean in cooperation with experts from Chalmers University of Technology, Gothenburg and the Royal Institute of Technology, Stockholm. The results of these tests confirmed the theoretical calculations. A hydrodynamic yield of 50 % was reached , i.e., 50 % of the energy content of the incoming waves was absorbed by the offshore wave energy converter. In an October storm, the converter (3 m diameter) generated peak values of 25 kW.

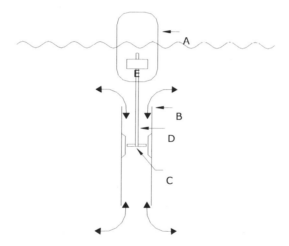

Figure 9.6 Principle of the IPS Buoy
Source: Bengt-Olov Sjöström, Chalmers Univ. of Technology, Sweden. (Reprinted by permission.)

[KEY The system consists of a circular/oval buoy (A), held in position by an elastic mooring system, enabling it to move up and down against a damping water mass contained in a long vertical tube (B) – the acceleration tube – underneath the buoy. The relative movement between the buoy and the water mass is transferred by the working piston (C) into the energy conversion system (D), which in turn drives the generator (E).]

9.3 Denmark

A formal Danish wave energy program was established in 1997 (Nielsen & Meyer 2000). The program is a result of a political agreement in 1996 to develop and promote new renewable sources of energy and storage options. The agreement includes financial support for wave energy research and development amounting to 40 mill. DKK over the years 1998, 1999, 2000 and 2001. The program is being carried out under the Danish Energy Agency. The European Union (EU) also provides support through its JOULE program. To advise on appropriate testing and research, the Danish Energy Agency has established an advisory panel of experts representing hydraulic and maritime institutes in Denmark, the Folkecenter for Renewable Energy, the University of Aalborg, the Technical University of Denmark and the Danish Wave Energy Association. The latter was formed in the spring of 1997 and has approximately 120 members. It can select and support new ideas with financial support (up to 50.000 DKK) for model constructions and testing; so far 35 new projects have been supported by the Association.

Advanced stage developments[3]

Before the formal Danish wave energy program was established, effort concentrated on the Danish Heaving Buoy offshore system originating from the research of Kim Nielsen, Technical University of Denmark, in the late 1970s. This system, which has been developed to an advanced stage, is based on the *freely heaving float with sea-floor reaction point*[2] wave energy conversion process. A 1 kW prototype of heaving and pitching floats was tested in 1985, and subsequently a circular buoy was developed that absorbs wave energy mainly from heave, and to a lesser extent from surge. Danish Wave Power Aps (DWP), a consortium of four Danish companies, was formed in 1989 to develop the device. During 1988-90, a 45 kW prototype was tested off Denmark's northwest coast, near the port of Hanstholm (Fig. 9.7) (Nielsen and Scholten 1990). Based on this work, a conceptual design for a 300 MW commercial plant was prepared, and cost and performance data were developed (Hagerman 1995a). A second sea test was carried out during 1994-1996 at the same location with a modified system. Based on these tests, the economic potential of full-scale wave power plants was revised, it being considered unlikely that wave energy would become commercially economic within the near future without a dedicated support structure for wave power development.

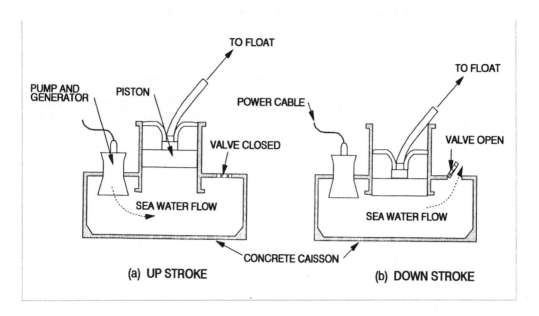

Figure 9.7 Danish wave power device
Source: Thorpe 1992 (Vol. 1). (Reprinted by permission.)

Other developments and activities

The Danish wave energy program is a bottom-up activity involving a broad selection of devices. In this regard it follows the same procedure as the successful Danish windmills program. In January, 2000, two years after the start of the formal program, work was underway on the following devices:

A.   DWP system (Danish Heaving Buoy system) *Danish Wave Power Aps. / RAMBØLL*, (Nielsen & Scholten1990) (see "Advanced stage developments", above, for details);

B.   Swan DK3 (based on the backward bent duct buoy) *Castlemain Scandinavia / Ralp Mogensen, Danish Hydraulic Institute*;

C.   Point absorber (development of point absorber with oil hydraulic power take-off), *RAMBØLL / Kim Nielsen*, (Nielsen & Smed 2000);

D.   Wave plane (pump without moving parts that utilizes both kinetic & potential energy), *Wave Plane International / Erik Skaarup*;

E.   Wave Dragon (floating wave energy converter of the overtopping type), *Lowenmark / Erik Friis Madsen, Aalborg University*, (Kofoed *et al.* 2000) (Fig. 9.8);

F.   Wave mill (horizontal axis), *LBHD v / Laurits M. Bernitt*;

G.   Wave turbine (vertical axis), *AUC / Tage Basse*;

H.   Wave pump, *Cambi A / S Ideutvikler Torger Tveter*;

I.   Wave plunger, *Danish Maritime Institute / Leif Wagner Smit*.

As of January 2002, the Danish wave energy program was terminated as a result of new political priorities set by the Danish government elected in November 2001. One project, the Wave Dragon, had obtained funding prior to this decision for building a large-scale (1:4.5) model of the device; additional funding for instrumentation and power take-off has been obtained from the European Union. Sea tests will be carried out at the sheltered Danish test site in Nisum Bredning in 2003.

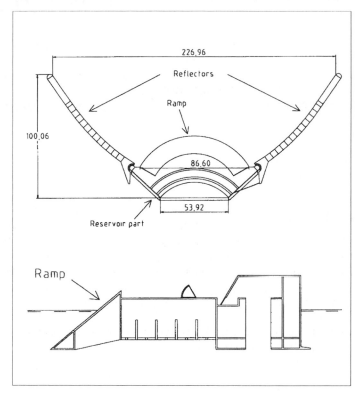

Figure 9.8 Wave Dragon. (Top – plan view; Bottom – cross section of reservoir)
Source: Kofoed *et al.* 2000

9.4 Germany

During the 1970s a German individual, Kayser (1974), proposed a submerged wave power device with hydraulic machinery. It was to be placed on the sea bed in shallow water, and it was to be driven by the hydrodynamic pressure. He also built a hand-driven model that he exhibited *inter alia* at the First Symposium on Wave Energy Utilization, Gothenburg, Sweden, in 1979.

Several years later, Graw (1993, 1994) at Bergische Universität GH Wuppertal studied a submerged plate device, and also an OWC. In the latter case he cooperated with the wave-energy team at ITT, Madras, India. More recently, he has been conducting wave energy research at the University of Leipzig (Graw *et al.* 2000).

9.5 Belgium

About twenty years ago, V. Ferdinande, State University of Ghent, conducted investigations on a floating point absorber (Ferdinande & Vantorre 1986). A heaving float, force-reacting against an essentially submerged body containing a large horizontal plate, was used for activating a piston pump (as opposed to the hose pump used with the similar Swedish wave power device). Other piston pumps in the mooring lines were envisaged to absorb energy from surge motion of the system.

9.6 The Netherlands

Advanced stage developments

The main activity in The Netherlands is focused on the Archimedes Wave Swing (AWS) device, which has progressed to an advanced stage of development. Based on the *submerged pulsating-volume body with sea-floor reaction point*[2] wave energy conversion process, this offshore device, based on an idea by Hans van Breugel, has been developed over the past several years by Teamwork Technology bv. Fred Gardner and Hans van Breugel are the key persons involved; the main shareholder of the company is NUON, a leading Dutch energy producer (utility). The original idea is described in International Patent Application (1995). The current device is described in the AWS Web Site (Archimedes Wave Swing 2001). The AWS consists of an upper part (the floater) of an underwater buoy, which moves up and down in the waves while the lower part (the basement or pontoon) stays in position. The periodic changing of pressure in a wave changes the buoyancy of the air contained in the floater, which initiates the movement of the upper part (Fig. 9.9). Hence, the floater is pushed down under a wave top and moves up under a wave trough. The power-take-off system consists of a linear electrical generator and a gas-filled damping cylinder. A successful feasibility study was undertaken in 1994. This was followed by Phase 1 of the project during 1995 and 1996 (with partners ECN and W.L. Delft Hydraulics), which involved the testing of a 1:20 scale model. Phase 2, during 1997 and 1998, involved the specification of a prototype plant together with model tests at both 1:20 and 1:50 scales. Based on these tests, it was agreed to develop a demonstration plant of 2 MW output (the commercial system will be in the 6 MW range). During

Phase 3 (1999-2001), the design of the demonstration plant was finalised, components ordered, and installation got underway. Future commercial systems are likely to comprise a number of inter-connected units (van Zanten 1996). It is planned to moor the demonstration plant off Viano do Castello, 100 km north of Oporto, Portugal. Plans call for this prototype to be tested during 2003.

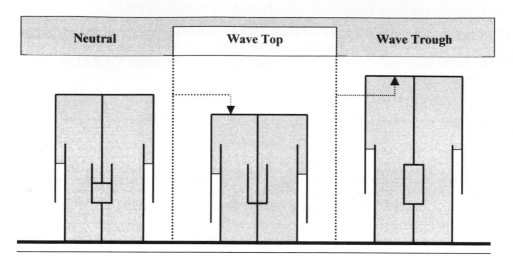

Figure 9.9 Principle of the Archimedes Wave Swing device
Source: Archimedes Wave Swing 2001. (Reprinted by permission.)

9.7 United Kingdom

In the 1970s, the UK had one of the largest government-sponsored R&D programs on wave energy, which covered a variety of devices (Davies 1985). However, this was significantly reduced in the early 1980s, and most recent work has centered on a shoreline oscillating water column (OWC) system developed at the Queen's University of Belfast (QUB) and built in a natural gully on the Isle of Islay, Scotland, in 1989 (Whittaker et al. 1991). In 1999, following a review of wave energy (Thorpe 1998) and other renewable energy sources, the Department of Trade and Industry (DTI) initiated a new program on wave energy, concentrating on those devices that had industrial support and were close to being tested as prototypes, e.g. LIMPET and Pelamis. In addition to DTI, wave energy research and development is funded by the UK Science and Engineering Research Council, as well as by electricity supply companies such as British Energy plc.

Following an extensive enquiry into the status and prospects for wave energy, the report of the House of Commons Science and Technology Committee on Wave Energy concluded that "The enormous potential export market for wave and tidal energy devices easily justifies the public investment now needed to ensure success" (UK Government 2000). As a result, several initiatives have followed.

For instance, in 2002 the UK Energy Minister announced plans for the development and demonstration of a series of new wave energy devices off the Western Isles of Scotland, including funding of up to £2.3 million for the British company Wavegen (Institute of Energy 2002b). Looking further ahead, a 2000 report issued by the Royal Commission on Environmental Pollution outlined various scenarios to achieve a 60% cut in carbon dioxide emissions in 2050. Several involved wave power and, depending on the scenario, it was stated that the year 2050 could see Britain having 7,500 wave power units, each of 1 MW (Royal Commission on Environmental Pollution 2000).

Operational wave energy devices[1]

Shoreline Gully, Isle of Islay (Fixed Oscillating Water Column[2]) A gully-based OWC was developed by researchers at the Queen's University of Belfast led by Trevor Whittaker as part of a long-term project sponsored by the U.K. Department of Trade and Industry (Whittaker & McIlwaine 1991; Whittaker & Raghunathan 1993). This led to a 75 kW prototype, which operated for several years on the Isle of Islay off Scotland's west coast (Fig. 9.10). While the onshore device performed well below its design capacity, it proved a valuable test facility, enabling study of the various components of the system under in-service conditions. After completing its purpose as a test facility, the device was decommissioned. The information and experience gathered by the team led to the development of a "designer gully" scheme known as the LIMPET.

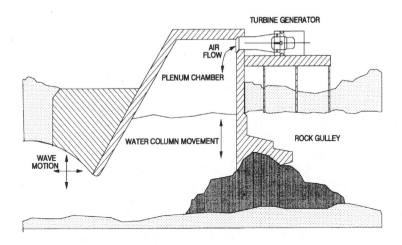

Figure 9.10 Shoreline gully OWC
Source: T.W. Thorpe, Energetech, Australia. (Reprinted by permission.)

LIMPET, Isle of Islay (Fixed Oscillating Water Column²): The LIMPET (Land Installed Marine Powered Energy Transformer) is a modular, shoreline-based oscillating water column developed by Wavegen of Inverness, Scotland (formerly Applied Research & Technology {ART}) in association with Queen's University of Belfast. It builds on the experience gained in the UK's only previous wave power device, the shoreline gully OWC on the Isle of Islay (see above) (Whittaker *et al.* 1996a, 1996b). The company successfully bid for a contract to include the device in the 1999 Scottish Renewables Obligation (SRO3). LIMPET follows the "designer gully" concept in which the device is constructed and fixed in place close to the shoreline, being protected from the sea by a rock bund. When the device is completely installed, the bund is removed, allowing the sea access to the device (Fig. 9.11). LIMPET consists of three water columns placed side-by-side in a man-made recess, which forms a slipway at an angle to the horizontal. In the current design for the Isle of Islay, the water column boxes are made from steel-reinforced concrete, giving a device width of ~21 m and a water plane area of 170 m². The device is anchored to rock promontories, its design being developed for ease of construction and installation, with minimal reliance on the existing coastline for suitable sites. In contrast to coastal wind-power schemes, LIMPET does not create a significant visual intrusion on account of its low profile. Power take-off is via two low-solidity, counter-rotating Wells' turbines, each rated at 250 kW. The device came into operation in September 2000, not far from the original prototype shoreline gully OWC (Wavegen 2000); the power plant has a 15-year agreement to supply electrical power to the major public electricity utilities (Institute of Energy 2001).

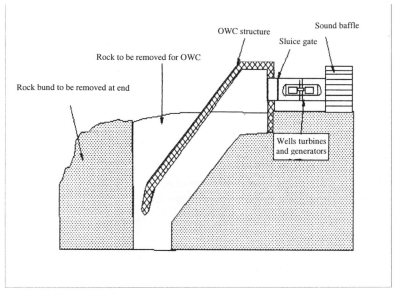

Figure 9.11 LIMPET OWC
Source: T.W. Thorpe, Energetech, Australia. (Reprinted by permission.)

110

Advanced stage developments[3]

Several UK wave energy devices have progressed to an advanced stage of development.

The most promising is the Pelamis, developed by Ocean Power Delivery Ltd. of Edinburgh, Scotland (Ocean Power Delivery 2000), and with support from the University of Edinburgh, this offshore device is based on the floating articulated cylinders with inertial reaction point[2] wave energy conversion process. The Pelamis device was conceived from the outset to use 100% "available" technology (i.e. all systems' components were to be available off-the-shelf, and the structure designed and fabricated to established offshore standards). It is a semi-submerged, articulated structure composed of cylindrical sections linked by hinged joints (Fig. 9.12). A novel joint configuration is used to induce a tuneable, cross-coupled resonant response that significantly increases power capture in light seas. Control of the restraint applied to the joints allows this resonant response to be "tuned-up" in light seas where capture efficiency must be maximized, or "tuned-down" to limit loads and motions in survival conditions. The wave-induced motion of the joints is resisted by hydraulic rams that pump high-pressure oil through hydraulic motors via smoothing accumulators. The hydraulic motors drive electrical generators to produce electricity. Power from all the joints is fed down a single umbilical cable to a junction on the sea bed. Several devices can be connected together and linked to shore through a single seabed cable. The complete device is flexibly moored so as to swing head-on to the incoming waves, and derives its reference from spanning successive wave crests. A 750 kW device will be 150 m long and 3.5 m in diameter. Among the key aspects of the device are the following:

- survivability conceived as the key objective;

- reacts against itself, rather than against a fixed reference frame such as the sea bed;

- incorporates features that inherently limit loads and motions once the rated-power wave amplitude has been reached;

- a degree of resonant response is introduced to improve power capture in small waves, with the device being de-tuned in large waves to prevent excessive loads and motions;

- the complete device will be constructed, assembled and tested off-site, with a minimum of installation work required on-site; and

- prototypes and initial production devices use 100% "available" technology, i.e. all components can be purchased off-the-shelf, with the structure being designed and fabricated to established offshore standards.

The company successfully bid for a contract to install a pair of 375 kW prototype devices off the Isle of Islay, Scotland, under the 1999 Scottish Renewables Obligation (SRO3). The devices are scheduled to be installed in 2003, and will generate over 2.5 million kWh of electrical power per year. The company also has a memorandum of understanding with BC Hydro for installation of a scheme in Vancouver Island, British Columbia, Canada (see Chapter 7, Section 7.2).

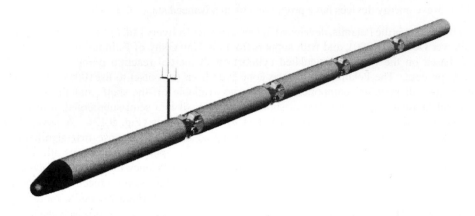

Figure 9.12 Pelamis device
Source: Ocean Power Delivery 2002. (Reprinted by permission.)

Another device is the near-shore OSPREY (Ocean Swell Powered Renewable Energy), is based on the *bottom-mounted oscillating water column*[2] principle. The original OSPREY design was for a hybrid wind/wave energy generation system. It was developed by Applied Research and Technology Ltd. (ART) of Inverness, Scotland (now Wavegen). Wave energy would be harnessed by means of an OWC within the inverted dome of a collector chamber between two large, pyramid-shaped ballast tanks which, when empty, enable the device to be self-floating for deployment (Fig. 9.13) (Hagerman 1996). As a result of the 1992 UK "Wave Energy Review" (Thorpe 1992), the OSPREY system underwent considerable development from the original domed-structure design to the final design that was launched in 1995. The development and construction of the device were supported by funds from industry and the European Commission (the JOULE II Programme). In particular, three important areas outlined in the UK Wave Energy Review (Thorpe 1992) received attention during this stage:

- *Capture Ratio.* This was improved to 116% (Thorpe 1995).

- *Structural Cost.* Conventional OWCs are massive concrete structures, the fabrication of which represents the largest capital cost item. These costs can be reduced by simplifying the device design and/or using alternative construction materials. As a result of these measures, civil engineering costs were reduced by over 30% compared to the earlier designs.

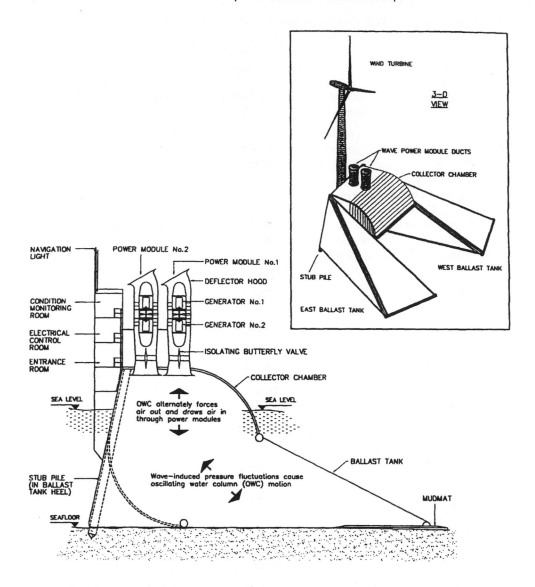

Figure 9.13 3-D view and general arrangement of OSPREY 1
Source: Hagerman 1996. (Reprinted by permission.)

- *Installation.* Most OWCs are bottom-mounted devices installed either in relatively shallow waters (<20 m depth) or on the shoreline. This normally requires preparation of the sea bed or shoreline, which (together with any schemes for anchoring the device) is usually the second largest cost item. Careful selection of suitable sites can reduce sea-bed costs and adoption of gravity-based anchoring can avoid expensive piling operations and simplify installation. As a result of these measures, transportation and installation costs were reduced by nearly 70% compared to the earlier designs.

The final prototype comprised a rectangular, steel collector chamber, which was open to the sea on one side. The oscillating water column had a power rating of 2 MW. The power module containing the turbines, generators, controls, etc. was mounted on top of the collector chamber, while hollow steel ballast tanks were fixed to either side. Behind the collector chamber and power module was a conning tower on which a "marinized" 500 kW wind turbine could be mounted. The device was designed for installation in a water depth of 14.5 m. It was intended that the device would be held in place by ballast contained in the two parts of the structure that protrude forward of the OWC collector. There were several aspects of the scheme that gave rise to some uncertainty with respect to the technical performance and integrity of the device and, while none were expected to be critical, financial pressure led to the design undergoing a rapid evolution. In August 1995, shortly after reaching its deployment site near Thurso, Scotland, the OSPREY failed structurally. It is thought that the device was damaged during launch, and that the affected area deteriorated during tow-out and installation, leading to the break up of the device. Following the failure, Wavegen redesigned many aspects of the scheme in order to ensure that the subsequent version, known as OSPREY 2000 (Fig. 9.14), would survive and function successfully (Thorpe 1998; Wavegen 2000). It was a requirement, for instance, that this version be a monolithic structure constructed in concrete. This stronger structure would also allow the installation of a larger wind turbine (1.5 MW rating or greater). The new Osprey would be of a composite construction, with procedures designed to minimise the time required for installation in open waters. It would be designed to operate in 15m of water within 1 km of the shore, generating up to 2MW of power for coastal consumers. Some key innovative features of the technology include:

- modular, low cost composite steel/concrete manufacture;

- rapid installation and decommissioning;

- minimal environmental impact; and

- 60-year structural design life with 20-year plant upgrades

An independent analysis (Thorpe1998) concluded that this system should produce electricity at a lower cost than any of the devices studied in the former UK Wave Energy Programme. As of 2000, the company had drawn-up plans for installing the first such device off the west coast of Ireland.

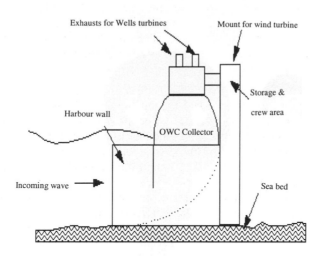

Exhausts for Wells turbines

Mount for wind turbine

Storage &

crew area

Harbour wall

OWC Collector

Incoming wave

Sea bed

Figure 9.14 OSPREY 2000
Source: T.W. Thorpe, Energetech, Australia. (Reprinted by permission.)

Also developed to an advanced stage is the SPERBOY offshore system, which is based on the *floating oscillating water column*[2] wave energy conversion process. A floating multiple resonant broadband Seapower energy recovery buoy, it has been developed by Embley Energy Ltd. (Fig.9.15). A pilot plant was installed in Plymouth Harbour in mid-2000. The project has received support from the European Commission.

The Contouring Raft offshore system is another UK device developed to an advanced stage . It is based on the *contouring float with sea-floor reaction point*[2] wave energy conversion process. In the mid-1970s, two contouring raft concepts with hydraulic power take-off were proposed independently and almost simultaneously: one in the United States by Glenn Hagen (refer to Chapter 7, Section 7.1, United States); and the other in the United Kingdom by Christopher Cockerell, inventor of the Hovercraft. This device was developed as part of the original UK wave energy program in the 1970s but further development work has ceased.

Figure 9.15 SPERBOY
Source: L. Bergdahl, Chalmers Univ. of Technology, Sweden. (Reprinted by permission.)

The final UK device discussed under the category of advanced stage developments is the SEA Clam. It is an offshore system utilizing the *flexible pressure device*[2] wave energy conversion process. Based on research and development at Coventry Polytechnic that got underway in 1978 as part of the UK's Wave Energy Programme, the spine-based, flexible-bag device was predicted to have one of the lowest electrical generating costs of any of the devices considered at that time (Davies 1985). Its stability and capture efficiency were enhanced by adopting a circular configuration. The resulting "Circular SEA Clam" (Fig. 9.16) was tested at 1/15th scale in Loch Ness (Sea Energy Associates 1986). The final design was for a floating, toroidal dodecagon, 60 m across and 8 m deep, with each side supporting rectangular air cells formed by flexible bags (Lockett 1991; Peatfield 1991). The air cells are maintained at an average pressure of 15 kPa, and connected to each other by a manifold. In this way, differential wave action around the device causes different cells to inflate or collapse, thereby pumping air around the torus. This air movement is converted into electricity by a Wells turbine and a generator mounted between the cells. Although an independent assessment identified the SEA Clam as one of the most promising offshore devices (Thorpe 1992), work on this concept had, as of 2000, all but ceased.

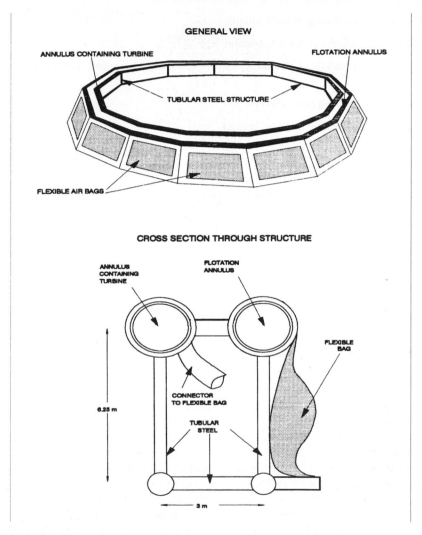

Figure 9.16 Circular Sea Clam
Source: Thorpe 1992 (Vol. 1). (Reprinted by permission.)

117

Other developments and activities

Numerous designs for wave energy devices were developed as part of the original UK wave energy programme (Dawson, 1979; Davies, 1985) but development work on most schemes ceased following the run down of that programme. The devices studied included:

1.   The Belfast Oscillating Water Column

2.   The Bristol Cylinder (see Ch.4, Section 4.1)

3.   The Cockerell Raft (see Contouring Raft System under Advanced stage developments, above)

4.   The Edinburgh Duck (see below)

5.   The Lancaster Flexible Bag (see below)

6.   The HRS Rectifier

7.   The NEL Floating Attenuator and Terminator

8.   The NEL Bottom Standing Terminator

9.   The SEA Clam (see Advanced stage developments, above)

10.   The Triplate

11.   The Vickers Submerged Terminator

12.   The Vickers Submerged Attenuator

Research on wave energy at Edinburgh University has been underway since the 1970s under the directorship of Stephen Salter (Salter 1974), the most famous example being the "Duck", an efficient but complex offshore device (Fig. 9.17) (Salter 1985). An assessment of the device identified areas requiring improvement (Thorpe 1992), and considerable work on these topics has been undertaken by the university. This has led to the design of a simpler, more economic system (Salter 1994b).

More recently, work at Edinburgh has expanded into other areas (University of Edinburgh, Wave Power Group 2000):
- Design and manufacture of a high efficiency, computer controlled, oil-hydraulic power take-off system (Salter & Rampen 1993).
- Design and manufacture of a variable-pitch turbine and high speed valve for the European Commission's pilot plant in the Azores (Salter & Taylor 1996) and a variable pitch Wells turbine for the same facility (refer also to Chapter 10, Section 10.1, Portugal).
- Work on a version of the IPS Buoy (refer to Ch.9, Section 9.2, Sweden) known as the "Sloped IPS Buoy" (Salter & Lin 1996). Controlling the angle of inclination of the buoy enables increases in the device's capture efficiency and bandwidth (compared to the original IPS

118

concept). An independent assessment of the buoy (Thorpe 1998) indicated that this was one of the most promising devices under development at that time.

- Building a new combined wave and tidal current test tank (University of Edinburgh, Wave Power Group 2000).

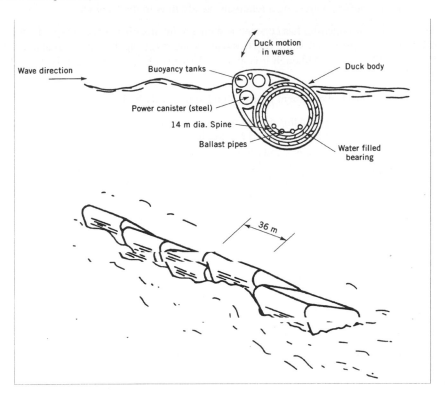

Figure 9.17 Nodding "Duck" concept developed at the University of Edinburgh. (The power conversion system is contained within sealed canisters and uses a gyroscopic inertial reaction point.)
Source: Thorpe 1992 (Vol. 1). (Reprinted by permission.)

Research and development at Lancaster University under French has concentrated on a diaphragm system, the Lancaster Flexible Bag, subsequently developed by Wavepower Limited in the United Kingdom (Platts 1982). The device consists of a straight, buoyant spine, 257 m in length, oriented perpendicular to the prevailing wave crests. A series of flexible bags is mounted on each side of the spine, manifolded into high- and low-pressure air ducts running along the spine's interior. The bags

119

are compressed as wave crests pass by, forcing air through non-return valves into the high-pressure duct. The bags expand during passage of wave troughs, withdrawing air from the low-pressure duct through another set of non-return valves. A conventional air turbine is mounted in the passage between high- and low-pressure ducts at the spine's end. The manifolded duct system thus acts as a short-term storage buffer, delivering a relatively steady flow to the turbine.

More recently, work at Lancaster has concentrated on a point absorber wave energy device, the PS Frog. This device comprises a floating flap that pitches and surges against an internal mass (French 1991; Thorpe 1992, Vol.1). Although it is at an early stage of development, this is a promising device.

Among UK Research & Development projects funded primarily by industry are:

A.   POWERBUOY. Work on this project was initiated by Wavegen in conjunction with the oil industry. It is an offshore, multi-MW, floating wave station for the supply of power to: satellite wellheads for pumps and other equipment; and power-deficient platforms in order to extend productive field life. The technology is primarily intended to be utilized in the development of marginal, offshore oilfields where deployment of a major production platform may not be economically justified.

B.   Floating Wave Power Vessel (FWPV). In 1999 it was announced by the UK's Scottish Office that Shawater Ltd. had been offered a supply contract to construct a 1.5 MW device in the Shetland Islands based on the Swedish FWPV concept (refer to Ch. 9, Section 9.2, Sweden).

C.   A reservoir-filling concept, proposed for the island nation of Mauritius by A.N. Walton Bott of the United Kingdom Crown Agents (Walton-Bott et al. 1988). The plan, which was developed by industry, involves the surging of waves up a ramped seawall, built along an existing reef that encloses a natural lagoon (which would act as the reservoir). The project has not been implemented

9.8 Ireland

At the national level, support of wave energy conversion activities in Ireland is provided by the Marine Institute and the Department of Energy. The principal centre for wave energy research is University College, Cork. Of particular note, the offshore McCabe Wave Pump has been designed to produce potable water and/or electricity.

Advanced stage developments[3]

The McCabe Wave Pump, which has been developed to an advanced stage, is based on the *contouring float with inertial reaction point*[2] wave energy conversion process (Fig. 9.18). The device was designed by Peter McCabe and a team of engineers from Hydam Technologies Ltd. in conjunction with Michael McCormick, The Johns Hopkins University, USA. The purpose of the device is to produce potable water (by reverse osmosis) and/or electricity for remote island communities, third-world countries with desert coastlines and remote military bases on the coast (McCormick *et al.* 1998). The McCabe Wave Pump uses a coupled-barge system in which wave-induced barge motions drive either open- or closed-hydraulic systems. After being devised by McCabe in 1980, it has been studied both theoretically and experimentally. In August 1996, a 40 m-long prototype was launched in the Shannon River estuary near Kilbaha, County Clare. The device consists of three rectangular steel pontoons, which are hinged together across their narrowest width (4 m). These pontoons are aligned so that their longitudinal direction heads into the incoming waves; they are moored using buoys and anchor chains. The three parts of the pontoon move relative to each other in the waves; the middle pontoon being held relatively still by a sub-sea damper plate, allowing the fore and aft pontoons to pitch about the hinges. Energy is extracted from the rotation about the hinge points by linear hydraulic pumps, mounted on the central pontoon near the hinges. This power can be taken off in two ways: (a) the pump pressurizes a closed-loop hydraulic oil system, which drives generators; or (b) the pump pressurizes an open-loop sea water system, which drives a Pelton turbine and generator (or alternatively pumps the seawater into a reverse-osmosis desalinator). During four months of prototype testing, the device functioned satisfactorily in swell heights of 1.0 to 2.5 m with a period of 7.5 s. The hydraulics dampen the pontoon pitching when the sea state exceeds 3 m wave height. The six pumps used on the system tested in 1996 (three on each power barge, and part of a closed oil hydraulic system) were undersized. As a result, the hydraulic system failed due to pressures in the system exceeding the limit of the hydraulic lines.

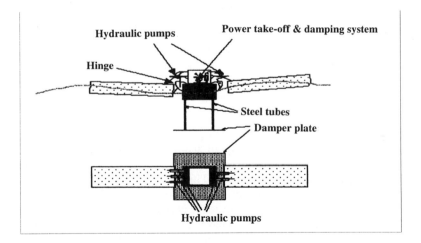

Figure 9.18 McCabe wave pump
Source: T.W. Thorpe, Energetech, Australia. (Reprinted by permission.)

Several modifications were subsequently incorporated, the most significant of these being:

- the center barge was made water-tight, and the beam increased to 5 m; and

- the hydraulic system was re-configured with a fail-safe system.

A prototype of the device is scheduled for deployment in 2003 in the Shannon Estuary, approximately 500 m offshore from Kilbaha (McCormick 2001). A 2001 estimate of the cost of potable water production using this device was $7 per 1000 US gallons (McCormick 2001).

Other developments and activities

The principal center for wave energy research in Ireland is University College, Cork, and activities there include the determination/evaluation of the wave resource, modeling the hydrodynamics of wave energy devices, model testing, and device design (primary OWCs). It has participated in many ongoing and completed projects funded by the European Commission, including participation in the European Wave Energy Atlas initiative. The university also coordinates the Wave Energy Research Network for the European Commission, and has conducted a number of confidential tests for private commercial companies.

The national Marine Institute has supported a number of studies of a generic nature related to the development of wave energy utilization under the auspices of the Marine Resources Measures program. The most significant has been a detailed resource study for the entire Irish coastline (undertaken by University College, Cork), which identified some 73 prime offshore sites where large numbers of wave energy generators could be anchored. The relevant government departments have addressed complementary issues, such as the licensing requirements for shoreline energy development and near-shore sea use. A second major study was an experimental/numerical study on the airflow across a Wells turbine (undertaken by the University of Limerick). Smaller, significant experimental investigations funded include: an experimental model study on the Rock OWC (see below); a similar study on the deQuesne device (see below); and a preliminary study on the BBDB (Backward Bent Duct Buoy) invented by Yoshio Masuda in Japan. These investigations were also conducted at University College, Cork.

The Rock OWC, or the Lindblom concept, is being pursued by a Swedish-Irish consortium under the lead of William Kingston at the University of Dublin. The concept involves blasting OWCs directly into steep coastal rocks by a low-cost method proposed by Prof. Lindblom at Chalmers University of Technology, Sweden, and confirmed in a bid by LKAB (a Swedish iron ore mining company). With an underground turbine, the only visible impact would be a small exhaust at the surface. If the technique works well, it will create the possibility of building many similar devices along stretches of the Irish coast. Model tests have been successfully undertaken in Cork, as well as theoretical work in Dublin and at Chalmers (Kingston *et al.* 2000).

A wave energy study, involving the Belfast engineering firm Harland & Wolf, the Marine Institute, and duQuesne Environmental Ltd. of Dublin was announced in July 1999. It is associated with a new wave energy conversion device developed by duQuesne. University College, Cork, Trinity College, Dublin, and Queen's University, Belfast are also involved (Ahlstrom 1999).

---

1. Category comprises full-scale devices, chiefly prototypes, that are currently operating (or have operated) where the energy output is utilized for the production of electricity or other purpose; also includes full-scale devices at an advanced stage of construction.

2. Refer to Table 3.1, Chapter 3, for details of the wave energy conversion process classification system used in this book; the system is a modified and updated version of that developed by George Hagerman (see Fig. 3.1, Chapter 3; also refer to Hagerman 1995a).

3. Category comprises: (a) devices of various scales, including full-scale, that have been deployed and tested *in situ* for generally short periods but where the energy output has not been utilized for the production of electricity or other purpose (in most cases plans call for the systems to be further developed and deployed as operational wave energy devices); and (b) full-scale devices planned for construction where the energy output will be utilized for the production of electricity or other purpose. Note: devices at an early stage of development are not included.

Chapter 10

WAVE POWER ACTIVITIES IN SOUTHERN EUROPE AND AFRICA

This geographic region has a long coastline but more modest wave exposure than the northern European region outlined in Chapter 9, e.g., 20 to 30 kW per meter of wave crest length off the west coast of Portugal. Levels in the Mediterranean are lower; for example, the exposed southern shoreline of Greece and its islands has levels of 5 to 15 kW per meter. In the case of Africa, the coasts of South Africa and South West Africa are exposed to the Westerlies, and exhibit wave power levels of 40 to 50 kW per meter of wave crest length.

Wave energy conversion activities in this geographic region are primarily based in three countries: Portugal, France and Greece. Some work has been undertaken in Spain, and a new initiative is getting underway in South Africa. In Portugal, where the emphasis is on the oscillating water column device, part of the work has been funded under the European Union's wave energy research and development program. This country has built a pilot plant on the island of Pico in the Azores that supplies electrical power to the island. France has a small research program based at Ecole Centrale de Nantes, which specializes in numerical simulation of the hydrodynamics of wave-body interaction. Together with an area off southern France, the wave power resources of Greece are considered the highest in the Mediterranean and exploitable. Various studies are underway in Greece, and plans are being developed, but no plants have yet been built. In Africa, investigations are underway in South Africa to evaluate if wave power is a viable electricity supply option.

10.1 Portugal

Research and development activity has been underway since 1978 at the Instituto Superior Tecnico (IST) of the Technical University of Lisbon, and since 1983 at INETI (National Institute of Engineering and Industrial Technology) of the Portugese Ministry of Economy. Most of the work is devoted to oscillating water column (OWC) systems, and includes the development of a pilot plant that is feeding electricity into the grid (Falcão 1999).

Operational wave energy devices[1]

Pico Power Plant, Azores (Fixed Oscillating Water Column[2]). A suitable site for this shoreline OWC was identified at Porto Cachorro on the northern coast of the island of Pico. Detailed bathymetric and topographic surveys, wave measurements and an energy resource evaluation were carried out for the site. The project is sponsored by the European Commission (EC) through the non-nuclear energy program JOULE, and coordinated by the Instituto Superior Técnico. Other partners are INETI, the Queen's University of Belfast (UK), the University College, Cork (Ireland), two Portuguese utilities (EDA and EDP), EFACEC (Portugal) and Profabril (Portugal). The bottom-standing concrete structure consists of a square platform (3.66 m x 12 m[2] inside dimensions), which spans a rocky gully where natural concentration of wave energy has been observed to occur (Fig. 10.1). The plant is equipped with a mono-plane Wells turbine (2.3 m diameter) driving a variable speed induction-type electrical generator (Fig. 10.2). Rated electrical power is 400 kW. The plant operated and supplied electrical energy to the island grid for the first time in October, 1999. Features of the plant include:

- Variable speed turbo-generator (750-1500 RPM), enabling efficient response to a wide range of sea states. It has a short-term smoothing effect (flywheel effect) on the electrical power supplied to the grid (Justino & Falcão 1999). The power electronics (incorporating a Kramer link), as well as the hardware and software for rotational speed control, were developed by partner EFACEC.

- Wells turbine with a double row of guide vanes (Gato & Falcão1990; Gato *et al.* 1996); unlike the UK LIMPET and Osprey turbines, which do not have such vanes.

- Inclusion of a relief valve in the roof of the air chamber structure for the purpose of suppressing large air pressure peaks, thus preventing pressure oscillations inside the chamber from exceeding the limit (depending on the instantaneous rotational speed) above which aerodynamic stalling of the turbine would produce a severe drop in power output. This measure also improves plant performance (Brito-Melo *et al.* 1996; Falcão & Justino1999).

As well as being a research and development facility, the plant is also designed as an operational power generation unit, permanently connected to the island grid to supply 8-9% of the annual electrical energy demand of the Island of Pico (population 15,000). The rated power output is about one third of the minimum night-time power demand of the island. This dual role was essential in ensuring financial support from the two electricity utilities. The mechanical and electrical equipment have specifications for a 25-year lifetime. Grid integration of the Pico plant has been one of the tasks of another European Commission JOULE project. A second Wells turbine, driving an identical generator, was manufactured and will be installed side-by-side with the plant's first set. The second turbine is of the variable-blade-pitch type. The electrical power conditioning system will be shared by the two sets. A high-speed stop valve, mounted in the turbine duct, will enable phase control by latching to be performed (Salter & Taylor 1996). Both the turbine and the valve were jointly designed and constructed by the Instituto Superior Técnico (Portugal) and the University of Edinburgh (UK), within the framework of a European Commission JOULE project.

Figures 10.1 (Top) Pico,Azores, wave-energy pilot plant – turbine exit at left & cable duct to grid at right; and 10.2 (Bottom) Pico plant – turbine & generator with part of ducting removed
Source: T. Pontes, INETI-Department of Renewable Energies, Portugal. (Reprinted by permission.)

127

Other developments and activities

Early Portugese research in the wave energy field included hydrodynamic modelling, both theoretical and experimental (in the wave basins and flumes of the National Civil Engineering Laboratory, Lisbon) (Sarmento 1993). More recently, boundary-element computer codes have been used in the hydrodynamic modeling of OWCs (Brito-Melo et al.1999, 2000). Other research includes theoretical and numerical modeling of oscillating-body, wave energy devices (floating and submerged), and theoretical studies on the hydrodynamic interaction among devices in arrays.

Research on Wells turbines has been ongoing since the early eighties, with the following configurations being studied, both theoretically and experimentally:

• turbines with monoplane and biplane rotors;

• contra-rotating turbines;

• variable-pitch turbines; and

• turbines with and without guide vanes.

Design methods and computer codes have been developed for the Wells turbine based on three-dimensional flow analysis. Aerodynamic testing has been performed at several rigs on 0.6 m diameter turbine models. The use of a variable-pitch Wells turbine as a means of phase-controlling an OWC, as well as improving the turbine efficiency, was proposed for the first time, and has been studied by theoretical simulation (wave-to-shaft energy conversion) and experimentally (Sarmento et al.1990; Gato & Falcão 1991; Perdigão 1998; Caldwell & Taylor 2000; Chatry 1999).

Wave-basin model testing methodology (Sarmento 1993), as well as theoretical/numerical simulation models (wave-to-electricity) (Brito-Melo et al. 1996) were developed at IST as a requirement for designing and building a full-size wave energy plant. Special attention has been devoted to device control studies and the development of plant control (including phase-control) strategies and procedures (valve control, rotational speed control, turbine blade-pitch control) (Perdigão 1998; Chatry 1999; Falcão & Justino 1999) .

In addition to national resource assessment studies, INETI coordinated a European Union initiative on the development of a common methodology for resource evaluation and characterization (Pontes et al. 1993). This led to the production of the European Wave Energy Atlas (Pontes 1998), which was also coordinated by INETI. The atlas incorporates a wide range of annual and seasonal wave-energy and wave-climate statistics covering the Atlantic and Mediterranean European coasts. Figure 2.8 (Chapter 2) presents the annual offshore gross power level and its directional distribution in the most energetic area of the northeastern Atlantic. In addition, the development of a mathematical model to compute the near-shore resource (Oliveira-Pires et al. 1996) and general resource studies for Europe (e.g. Oliveira-Pires et al. 1999) have been carried out. This enabled the development by INETI of ONDATLAS, an atlas of the nearshore wave climate and resource in Portugal (Pontes et al. 2001).

10.2 Spain

In 1990, the Unión Eléctrica Fenosa of Spain commenced research on a wave energy scheme involving an oscillating water column device where the power is extracted, not by an air driven turbine, but mechanically using a float on top of the water column. An experimental plant was tested in a wave flume, allowing its behaviour to be modelled. Following further testing, plans were drawn-up to install a prototype in a breakwater.

10.3 France

Preliminary design work on an artificial beach-wall, wave power system for Maré Island (New Caledonia) was undertaken in the early 1980s by IFREMER and the Ecole Centrale de Nantes (ECN). IFREMER[3] is the French public research institute for oceanic and maritime matters (at that time it was named CNEXO). ECN[4] (at that time named ENSM) is an academic institute of technology based in Nantes and devoted to engineering in a range of specialties, including ocean engineering, as well as research in several laboratories, including one of naval architecture and ocean engineering[5] A thesis project based on the Maré Island wave power project (Spiridakis 1983) resulted in three publications (Clément & Spiridakas 1983; 1984a; &1984b). Unfortunately the funding ceased in 1984, and work on the project was discontinued.

From 1985 to 1995, theoretical/numerical studies were undertaken by A. Clément's wave energy group at ECN (Nantes), indirectly funded through students' grants. During this period, research was reoriented towards the dynamic absorption problem, namely the development of wave absorbing devices (paddles) for the equipment of wave basins (Maisondieu 1993).

In 1995, A. Clément and A.J.N.A Sarmento (IST[6] Lisbon, Portugal) established collaborative studies between their respective institutes on (sub)-optimal control strategies for wave energy oscillating water column (OWC) devices. These studies were based on IST's involvement in the European wave power pilot plant on the island of Pico (Azores) (refer to Section 10.1, Portugal) and ECN expertise in numerical simulation of the hydrodynamics of wave-body interaction. Two common ("co-tutelle") PhD thesis projects arising from the collaboration were undertaken and resulted in several publications (Clément 1997; Brito e Melo *et al*. 2000; Chatry *et al* 1999, 2000; Brito e Melo *et al*. 1999; Brito e Melo *et al*. 2000).The main practical results of this collaboration are:

A.   a computer code (AQUADYN-OWC) that simulates the hydrodynamics of OWC wave power plants, and accounts for the influence of surrounding bathymetry (this parameter was shown to be of major importance);

B.   a methodology for the time-domain simulation of the whole process, from the waves to the electric current delivered to the grid; and

C.   a self-adaptive control strategy ("SAFF") for the online control of wave power devices.

The theoretical work, still indirectly funded via student grants, continues at ECN. It includes the development of a computer program devoted to the time-domain simulation of 3D OWC wave power plants in non-linear hydrodynamic theory.

10.4 Greece

The exploitation of wave energy in Greece has not attracted significant industry involvement because the annual wave power level in the seas around Greece is moderate (typically 5-10 kW/m). This is less than in the open seas of the Atlantic Ocean, where the annual wave power average often exceeds 40-50 kW/m. There is, however, certainty for the presence of "hot spots" in the Aegean Sea caused by the complex island terrain, but these have not yet been investigated in detail. In such locations, the annual wave power level would be of the order of 15-20 kW/m. Together with an area in southern France, the wave power resources in Greece are considered to be among the highest in the Mediterranean and exploitable. An important reason for this is that, being moderate, the wave climate does not require high safety factors to be built into the design of wave power plants, thus implying lower construction and investment costs.

In 1996, preliminary plans were developed for a full-scale, semi-commercial demonstration plant for fresh water and electricity production on the island of Amorgos in the South Aegean Sea. These were based on the Hosepump and/or the IPS buoy (refer to Ch.9, Section 9.2, Sweden), and arose out of the joint efforts of IPS (Sweden), Technocean (Sweden), Dunlop Oil & Marine (UK) and Sercon (Greece) (Sjöström, Economides & Gallant 1996). The plans called for high-basin energy storage (a mixture of the Hosepump and Tapchan devices).

In 1995, the European Commission sponsored a project to develop a novel oscillating water column design for mounting in a breakwater, where the output from several units would be manifolded together to drive air pistons.

Studies in wave power modelling are currently (i.e. 2000) being carried out by G.A. Athanassoulis and his group at the Laboratory of Ship and Marine Hydrodynamics (Dept. of Naval Architecture and Marine Engineering), National Technical University, Athens. These include offshore, near shore and onshore studies (Athanassoulis & Belibasssakis 1998; Athanassoulis & Belibasssakis 1999; Athanassoulis et al. 1996; Cavaleri et al. 1999).

Among wave measurement activities in Greek waters, is work under the international program POSEIDON. Carried out by NCMR (National Centre for Marine Research) and OCEANOR (Norway) under the responsibility of G. Chronis/T. Soukissian, the project involves a network of eight observation buoys continuously recording physical and other parameters of the Greek seas. The data are transmitted to an operational centre where they are fed into forecasting models. The system has been in operation since 1999 (POSEIDON 2001).

The development of models of interacting point absorbers has been undertaken by S. Mavrakos, Division of Offshore Structures (Dept. of Naval Architecture and Marine Engineering), National Technical University, Athens (Mavrakos & Kalofonos 1996; Mavrakos & Kalofonos 1997;

Mavrakos & McIver 1997). The work was funded by the European Commission under the program "Offshore Wave Energy Converters", OWEC-1.

10.5 South Africa

EsKom, South Africa's largest power producer, has embarked on an investigation to evaluate if utility-scale, renewable energy, including wave power, is a viable supply-side option for EsKom and South Africa. The overall program, known as the South African Bulk Renewable Energy Program (SABRE-Gen), was initiated in 1998 (SABRE-Gen 2002). In addition to wave power, it covers solar, biomass and wind energy. The first phase of the wave power component was a resource assessment of the wave power potential along South Africa's coastlines. The study concluded that the resource available is sufficient, with wave power levels well in excess of 30 kW/m. The second phase, currently underway (2002), involves a techno-economic and technology selection study.

----

1. Category comprises full-scale devices, chiefly prototypes, that are currently operating (or have operated) where the energy output is utilized for the production of electricity or other purpose; also includes full-scale devices at an advanced stage of construction.

2. Refer to Table 3.1, Chapter 3, for details of the wave energy conversion process classification system used in this book; the system is a modified and updated version of that developed by George Hagerman (see Fig. 3.1, Chapter 3; also refer to Hagerman 1995a).

3. Institut Français de Recherche pour l'Exploitation de la mer

4. Ecole Centrale de Nantes, 1 Rue de la Noe, 44300 Nantes, France

5. Laboratoire de Mécanique des Fluides, Division Hydrodynamique Navale

6. Instituto Superior Tecnico, Av. Rovisco Pais, Lisbon, Portugal

Chapter 11

CONCLUSIONS AND FUTURE PROSPECTS

Wave energy utilization is a technology that is still at a very early stage of development. In certain niches of the market it is commercially competitive, such as in the supply of power for navigation buoys, in the wave-powered pumping of water for desalination plants, and in the generation of electricity for isolated coastal communities that currently depend on diesel generators. However, further innovation and technological development is required before the utilization of wave energy can be introduced on a large scale in the general energy market.

It is a well known observation that, due to experience and improved methods of production, the unit cost of a product usually diminishes as the production volume increases. A typical trend is a reduction of 20 to 25 percent of the inflation-corrected price for each doubling of the cumulative production (Fischer 1974). For this reason, there are good hopes for the large-scale utilization of wave energy in the future, since the energy costs of the best present schemes are not excessively high compared to current market price (Falnes 1996).

The cost reduction due to experience and innovation illustrates the handicap that new energy technologies face in market competition with well-established, conventional energy technologies. This observation should be borne in mind when comparisons are made between the energy costs of new technologies and the energy costs of conventional plants. As a human has to grow from conception to an adult person, so a new energy production method has to develop from an idea to mature technology. Using this analogy, we may perhaps say that wave energy is still in its infancy, wind energy is a teenager and conventional energy is an adult.

In some wave energy research programs, e.g., United Kingdom in the 1980s, it was assumed as a design goal that the devices should convert as large a fraction as possible of the wave energy impinging on a coast line. However, since the natural energy in the ocean is "free", this is not necessarily the best strategy. Instead, the installed power capacity, relative to the available power in the sea, should, in the future, be left as an open parameter in the economical optimization of wave energy converters. While such a strategy may result in reduced overall power production, it should achieve a higher duty factor and lead to better overall economic prospects. Also there will be less requirement for primary energy-storage capacity to even-out the effect of wave variability. To date, such considerations have not been the dominating strategy in the design of the majority of wave energy converters deployedin the ocean and along shorelines. Instead, rather simple technologies have been utilized, with many of the devices being essentially civil engineering structures that are probably too large to maximize the ratio between energy production and investment. To increase this ratio, more sophisticated designs are required.

Wave-energy devices of a simple type may offer better economic prospects if constructed in non-industrialized countries where labour is inexpensive. For this reason, wave energy may become commercial earlier in such countries than in industrialized countries. Also, in progressing to more advanced technology it will be necessary to develop new components, e.g. electronic hardware and software, pumps, valves, turbines, pneumatic and hydraulic motors. Such advanced technology will likely require local expertise for the operation and maintenance of installed wave energy converters.

A large offshore or near-shore wave energy plant can be envisaged as a collection of many (hundreds or thousands) primary wave-energy converting units, each with a power capacity in the range from 100 kW to 1 MW. Thus the production of such units will be serialized, which will reduce costs. Hydraulic or pneumatic energy can be collected from many such units and fed to a central unit containing a large turbine and electrical generator. This housing may be placed on the sea bed or, in the case of near-shore wave energy converters, on land.

REFERENCES CITED

ABB Ltd. 2002 <http://www.abb.se>

Ahlstrom, D. 1999 Wood sees bright future for energy from waves. *The Irish Times*, Jul. 15, 1999.

Allender, J., Audunson, T., Barstow, S.F., Bjerken, S., Krogstad, H.E., Steinbakke, P., Vartdal, L., Borgman, L.E. and Graham, C. 1989 The WADIC project, a comprehensive field evaluation evaluation of directional wave instrumentation', *Ocean Eng.,* **16**, 505-536.

Ambli, N., Budal, K., Falnes, J. and Sørenssen, A. 1977 Wave power conversion by a row of optimally operated buoys. *10th World Energy Conference*, Istanbul, Sept 19-23, 1977. Paper 4.5-2 (16 pages).

Ambli, N., Bønke, K., Malmo, O. and Reitan, A. 1982 The Kvaerner multiresonant OWC. In *Proc. 2nd Int. Symposium on Wave Energy Utilisation* (ed H. Berge), 275-295. Trondheim, Norway: Tapir. (ISBN 82-519-0478-1).

Anderssen, P.1999 Wave power activities in Indonesia. *Private Communication*: Per Anderssen, NORWAVE, N-1412 Sofiemyr, Norway.

AquaEnergy Group. 2002 < http://www.aquaenergygroup.com>

Archimedes Wave Swing. 2001 <http://www.waveswing.com>

Ardhendu, G.P., Jayashankar, V. and Ravindran, M. 1997 Performance of a Wells type turbine coupled to an induction generator for wave energy applications. In *Proceedings of theSeventh International Offshore and Polar Engineering Conference, Honolulu, USA, May, 1997,* **1**, 962-965. The International Society of Offshore and Polar Engineers.

Athanassoulis, G.A. and Belibassakis, K.A. 1998 Water wave green function for a 3Duneven-bottom problem with different depths at -oo and +oo. In *Proc. IUTAM Symposium on Computational Methods in Unbounded Domains, Boulder Colorado,USA, 27-31 Jul., 1997,* Int. Union of Theoretical and Applied Mechanics, Fluid Mechanics and its Applications **49**, 21-32. Kluwer Academics Publishers.

Athanassoulis, G.A. and Belibassakis, K.A.. 1999 A consistent coupled-mode theory for the propagation of small-amplitude water waves over variable bathymetry regions. *Journal of Fluid Mechanics* **389**, 275-301.

Athanassoulis, G.A., Pontes, M.T., Tsoulos, L., Nakos, B., Stefanakos, Ch.N.,Skopeliti, A. and Frutuoso, R. 1996 European Wave Energy Atlas: An Interactive PC-based system. In *Proc.*

2nd. *European Wave Power Conference, Lisbon, Portugal, 8-10 November, 1995* (eds. G. Elliot & K. Diamantaras). Brussels: European Commission. (ISBN 92-827-7492-9).

Athanassoulis, G.A. and Skarsoulis, E.K. 1992a *Wind and wave atlas of the North-Eastern Mediterranean Sea*. Athens, Greece: Laboratory of Ship and Marine Hydrodynamics, Nat. Tech. Univ.

Athanassoulis, G.A. and Skarsoulis, E.K. 1992b Wave climate and gross wave power estimates for the Northeastern Mediterranean Sea based on visual wave data. *Report, Nat. Tech. Univ., Athens, Greece.*

Bascom, W. 1980 *Waves and Beaches*, 2nd edition. Garden City, NY: Anchor/Doubleday

Barstow, S.F. and Falnes, J. 1996 *Ocean wave energy in the South Pacific, the resource and its utilisation*. South Pacific Applied Geoscience Commission (SOPAC) Miscellaneous Reports 234. (ISBN 982-207-002-0).

Bates, J. 1995 Full fuel cycle atmospheric emissions and global warming impacts. In *UK Electricity Generation, ETSU Report R-88*. Harwell, UK: Department of Trade & Industry.

BCHydro. 2001 <http://www.bchydro.com>

Beattie, W.C., Sprevak, D. and Alcorn, R. 2000 Producing acceptable electrical supply quality from a wave-power station. In *Proc. 3rd European Wave Power Conf., Patras, Greece, 30 Sep.-2 Oct., 1998* (ed. W. Dursthoff), 252-257. University of Hannover.

Bønke, K., and Ambli, N.1987 Prototype wave power stations in Norway. In *Utilization of Ocean Waves - Wave to Energy Conversion* (eds M.E. McCormick and C.K. Young), 34-44. New York, New York: American Society of Civil Engineers.

Brito e Melo, A., Hofmann, T., Sarmento, A.J.N.A., Clément, A.H. and Delhommeau, G. 2000 Numerical modelling of OWC-shoreline devices with the effects of surrounding coastline and non-flat bottom. In *Proc. 10th Int. Offshore & Polar Engng. Conf., ISOPE2000, Seattle.*

Brito-Melo, A., Sarmento A.J.N.A. and Gato, L.M.C. 1996 Mathematical extrapolation of tank testing results: application to the Azores Wave Pilot Plant. In *Proc. 2nd European Wave Power Conf., Lisbon, Portugal, 8-10 Nov., 1995* (eds. G. Elliot & K. Diamantaras), 141-147. Brussels: European Commission. (ISBN 92-827-7492-9).

Brito e Melo, A., Sarmento, A.J.N.A., Clément, A.H. and Delhommeau, G. 1999 A 3D boundary element code for the analysis of OWC wave-power plants. In *Proc. 9th Int. Offshore & Polar Engng. Conf., ISOPE99, Brest*. **I**, 188-195. (ISBN 1-880653-40-0).

Brito e Melo, A., Sarmento, A.J.N.A., Clément, A.H. and Delhommeau. G. 2000 Hydrodynamic analysis of geometrical design parameters of oscillating water column devices. In *Proc. 3rd European Wave Power Conference, Patras, Greece, 30 Sep.-2 Oct., 1998* (ed. W. Dursthoff). University of Hannover.

Budal, K. 1977 Theory for absorption of wave power by a system of interacting bodies. *Journal of Ship Research*, **21** (4), 248-253.

Budal, K. and Falnes, J. 1975 A resonant point absorber of ocean-wave power. *Nature*, 256, 478-479. (Corrigendum in **257**, 626).

Budal, K. and Falnes, J. 1977 Optimum operation of improved wave-power converter. *Marine Science Communications*, **3**, 133-150.

Budal, K. and Falnes, J. 1980 Interacting point absorbers with controlled motion. In *Power From Sea Waves* (B. Count, ed.), 381-399. London : Academic Press.

Budal, K., Falnes, J., Hals, T., Iversen, L.C. and Onshus, T. 1981 Model experiment with a phase controlled point absorber. In *Proc. 2nd Int. Symposium on Wave and Tidal Energy, Cambridge, UK, 23-25 Sep., 1981*, 191-206. Cranford, UK: BHRA Fluid Engineering. (ISBN 0-906085-43-9).

Budal, K., Falnes, J., Iversen, L.C., Lillebekken, P.M., Oltedal, G., Hals, T., Onshus, T. and Høy, A.S. 1982 The Norwegian wave-power buoy project. In *Proc. 2nd Int. Symposium on Wave Energy Utilisation* (ed H. Berge), 323-344. Trondheim, Norway: Tapir. (ISBN 82-519-0478-1).

Caldwell, N. and Taylor, J.R.M. 2000 An eddy current actuator for changing the pitch of turbine blades. In *Proc. 3rd. European Wave Power Conference, Patras, Greece, 30 Sep.- 2 Oct., 1998* (ed. W.Dursthoff). University of Hannover.

Cavaleri, L., Athanassoulis, G.A. and Barstow, S. 1999 Eurowaves: a user-friendly approach to the evaluation of near-shore wave conditions. In *Proc. 9th. International Offshore and Polar Engineering Conf. & Exhibition, ISOPE 99, Brest , France, 30 May - 4 June 1999*.

Chatry, G. 1999 Development et Simulation d'une Méthode de Régulation Auto-Adaptative pour l'Absorption Dynamique des Ondes de Gravité. *Doctor Degree*, Ecole Centrale de Nantes, Nantes, France and Instituto Superior Técnico, Technical University of Lisbon, Portugal.

Chatry, A.H., Clément, A.H. and Sarmento, A.J.N.A. 2000 Self-adaptive control of an OWC device. In *Proc. 3rd. European Wave Energy Conference, Patras, Greece, 30 Sep.-2 Oct , 1998* (ed. W. Dursthoff). University of Hannover.

Chatry, A.H., Clément, A.H. and Sarmento, A.J.N.A. 1999 Simulation of a self-adaptively controlled OWC in a non-linear numerical wave tank. In *Proc. 9th Int. Offshore & Polar Engng. Conf., ISOPE99, Brest*. **I**, 290-296. (ISBN 1-880653-40-0).

Cho, Kyu-Bock and Shim, Hyun-Jin. 1999 Development of a new wave energy generating system. In *Recent Advances in Marine Science and Technology '98* (ed. N.K. Saxena), 165-172. Honolulu, Hawaii: PACON International.

Claeson, L., Forsberg, J., Rylander, A. and Rindby, T. 1987 *Energi från havets vågor* (Energy from the waves of the ocean). Energiforskningsnämnden, Efn-rapport nr 21, Allmänna förlaget, Stockholm, Sweden. (ISBN 91-38-09691-9)

Claeson, L. and Sjöström, B.-O. 1998 Syresättning av bottenvatten med vågenergi (Aeration of bottom water with wave power). *Report 1988:134*, Göteborg, Sweden: Technocean.

Clément, A.H. 1997 Dynamic non-linear response of OWC wave energy devices. *Int. J. Offshore Polar Engng.* **7** (2), 154-159. (ISBN 1053-5381).

Clément, A. and Maisondieu, C. 1994 Comparison of time-domain control laws for a piston wave absorber. In *1993 European Wave Energy Symposium. Proceedings of an International Symposium held in Edinburgh, Scotland, 21-24 July 1993*, 117-122. (ISBN 0-903640-84-8).

Clément, A. and Spiridakis, M. 1983 Etude expérimentale de la récupération de l'énergie de la houle par les systèmes à rampe de déferlement. *Session de l'Association Technique Maritime et Aéronautique (ATMA).*

Clément, A. and Spiridakis, M.1984a Récupération de l'énergie de la houle par déferlement sur des structures trièdres. *Session de l'Association Technique Maritime et Aéronautique (ATMA).*

Clément, A. and Spiridakis, M. 1984b Structures trièdres côtières pour la récupération de l'énergie de la houle. In *Proc. Int. Symposium on Maritime Structures in the Mediterranean Sea, Athènes.*

Czitrom, S.P.R. 1997 Wave-energy driven resonant sea-water pump. *J. Offshore Mechanics & Arctic Engineering* **119**, 191-195.

Czitrom, S.P.R., Godoy, R., Prado,E., Pérez, P and Peralta-Fabi, R. C. 2000a Hydrodynamics of an oscillating water column seawater pump, Part I: theoretical aspects. *Ocean Engineering*, **27**, 1181-1198.

Czitrom, S.P.R., Godoy, R., Prado, E., Olvera, A. and Stern C., 2000b Hydrodynamics of an oscillating water column seawater pump, Part II: tuning to monochromatic waves *Ocean Engineering*, **27**, 1199-1219.

Dawson, J.K. 1979 Wave energy. *Energy Paper No. 42.* London, UK: HMSO.

Davies, C.G., Cotton, P.D., Challenor, P.G. and Carter, D.J. 1998 On the measurement of wave period from radar altimeters: ocean wave measurements and analysis. In *Proc. 3rd Int. Symposium on Waves, Reston, VA, USA*, 819-826. New York, New York: American Society of Civil Engineers.

Davies, P.G.(ed.). 1985 Wave energy - - the Department of Energy's R & D Programme, 1974-1983. *ETSU Report R-26.* Harwell, UK: Department of Energy.

Department of Physics, Norwegian University of Science & Technology. 2000 <http://www.phys.ntnu.no/instdef/grupper/miljofysikk/bolgeforsk/index-e.html>

Department of Trade and Industry. 1996 *Digest of United Kingdom Energy Statistics*. London: HMSO.

Dijk, J. van. 1992 Hydraulics with a little extra. *European Oil Hydraulics and Pneumatics*. Feb. 1992, 31-34.

Earle, M.D. and Bishop, J.M. 1984 *A practical guide to ocean wave measurements and analysis.* USA: Endeco Inc. 78p.

Edinburgh-Scopa-Laing Wave Energy Group. 1979 Wave energy-Salter's Duck fifth year report. *ETSU Report*, WV-1512, Parts 4A-4C. Harwell, UK: Department of Trade& Industry.

Eidsmoen, H. 1996. Optimum control of a floating wave-energy converter with restricted amplitude. *Journal of Offshore Mechanics and Arctic Engineering*, **118**, 96-102.

Eidsmoen, H. 1998. Tight-moored amplitude-limited heaving-buoy wave-energy converter with phase control. *Applied Ocean Research*, **20** (3), 157-161.

Electric Power Research Institute. 1987 Technical assessment guide, Vol. 3, Fundamentals and methods, supply-1986. *Report P4463-SR*. Palo Alto, California: Electric Power Research Institute.

Electric Power Research Institute. 1993 Technical assessment guide, Vol. 1, electricity supply-1993 (Revision 7). *Report TR-102276-V1R7*. Palo Alto, California: Electric Power Research Institute.

Energetech Australia Pty. Ltd. 2000 <http://www.energetech.com.au>

Eshan, M. *et al.* 1996 Simulation and dynamic response of computer controlled digital hydraulic pump/motor use in wave energy power conversions. *Proc. 2ⁿᵈ European Wave Power Conf., Lisbon, Portugal, 8-10 Nov, 1995* (eds. G. Elliot & K. Diamantaras), 305-311. Brussels: European Commission. (ISBN 92-827-7492-9).

European Commission. 1993 Wave energy converters: generic technical evaluation study. *Danish Wave Power APS*.

European Commission. 1994 An assessment of renewable energy technologies for the UK. *ETSU Report R-82*. London: Department of Trade & Industry and HMSO.

European Commission. 1995a DGXII, Science, Research and Development, JOULE (1995). Externalities of Energy, 'ExternE' Project, **1**, *Summary Report, No. EUR 16520 EN*.

European Commission. 1995b DGXII, Science, Research and Development, JOULE. Externalities of Energy, 'ExternE' Project, Vol. 4, Oil and Gas. *Report No. EUR 16523 EN, Part II, The Natural Gas Fuel Cycle*.

Evans, D.V. 1976 A theory for wave-power absorption by oscillating bodies. *Journal of Fluid Mechanics*, **77**, 1-25.

Evans, D.V. 1979 Some theoretical aspects of three-dimensional wave energy absorbers. In *Proceedings of First Symposium Ocean Wave Energy Utilization, Gothenburg, Sweden*, 77-113.

Evans, D.V. 1981 Maximum wave-power absorption under motion constrains. *Applied Ocean Research*, **3** (4), 200-203.

Evans, D.V. 1982 Wave-power absorption by systems of oscillating surface pressure distributions. *Journal of Fluid Mechanics*, **114**, 481-499.

Falcão, A. F. de O. 1999 Design and construction of the OWC wave power plant at the Azores. In *Proc. International One-Day Seminar on Wave Power: Moving Towards Commercial Viability*. London, UK: Institution of Mechanical Engineers.

Falcão, A.F. de O. and Justino, P.A.P. 1999 OWC wave energy devices with air flow control. *Ocean Engineering*, **26**, 1275-1295.

Falcão, A.F. de O. and Sarmento, A.J.N.A. 1980 Wave generation by a periodic surface pressure and its application in wave-energy extraction. In *Proc. 15th Int. Cong. Theor. Appl. Mech.* Toronto.

Falnes, J. 1980 Radiation impedance matrix and optimum power absorption for interacting oscillators in surface waves, *Applied Ocean Research*, **2** (2), 75-80.

Falnes, J. 1993 Research and development in ocean-wave energy in Norway. In *Proc. International Symposium on Ocean Energy Development, August 1993, Muroran, Hokkaido, Japan* (ed. H. Kondo), 27-39. (ISBN 4-906457-01-0).

Falnes, J. 1994 Small is beautiful: How to make wave energy economic. In *1993 European Wave Energy Symposium. Proceedings of an International Symposium held in Edinburgh, Scotland, 21-24 July 1993*, 367-372. (ISBN 0-903640-84-8).

Falnes, J. 1996 A crude estimate of benefit-cost ratio for utilization of the wave energy of the world's oceans. In *Proc. 2nd European Wave Power Conference, Lisbon, Portugal, 8-10 Nov. 1995* (eds. G. Elliot & K. Diamantaras), 102-104. Brussels: European Commission. (ISBN 92-827-7492-9).

Falnes, J. 2000 Maximum wave-energy absorption by oscillating systems consisting of bodies and water columns with restricted or unrestricted amplitudes. In *Proc. of The Tenth (2000) Int. Offshore & Polar Engineering Conf., 28 May-2 June, 2000, Seattle, USA*, **1**, 420-426. Cupertino, CA, USA: Int. Soc. of Offshore & Polar Engineers.

Falnes, J. 2002 Optimum control of oscillation of wave-energy converters. *Int. J. Offshore & Polar Engineering*, **12** (2), 147-155.

Falnes, J. and Budal K. 1982 Wave-power absorption by parallel rows of interacting oscillating bodies. *Applied Ocean Research*, **4** (4), 194-207.

Falnes, J. and McIver, P. 1985 Surface wave interactions with systems of oscillating bodies and pressure distributions. *Applied Ocean Research*, **7** (4), 225-234.

Ferdinande, V. and Vantorre, M. 1986 The concept of a bipartite point absorber. In *Proc. IUTAM Symposium on Hydrodynamics of Ocean Wave-energy Utilization, Lisbon, Portugal, 8-11 July, 1985* (eds. D.V. Evans & A.F. de Falcao), 217-226. Springer-Verlag.

Fernandes, A.C. 1985 Reciprocity relations for the analysis of floating pneumatic bodies with application to wave power absorption. In *Proc. 4th International Offshore and Arctic Engineering Symposium*, **1**, 725-730.

Fischer, J.C. 1974 *Energy Crisis in Perspective*. New York, NY: John Wiley.

Forristal, G.Z. 2002 Wave crest sensor intercomparison study: An overview of WACSIS. In *Proc. 21st International Conf. on Offshore Mechanics and Arctic Engineering*, Oslo, Norway, **28**, 448. OMAE.

Fredrikson, G. 1992 IPS wave power buoy, Mark IV. In *Proceedings of a Workshop on Wave Energy R & D, 1-2 October, 1992, Cork, Ireland*; *Report EUR 15079 EN*. Commission of European Communities.

French, M.J. 1991 Latest developments in wave energy at Lancaster. In *Proceedings I. Mech. E. Seminar on Wave Energy, 28 Nov., 1991*.

Gato, L.M.C. and Curren, R. 1997 The energy conservation performance of several types of Wells turbine designs. *Proc. I. Mech. E., Part A, J. Power Engineering*, **211** (A2), 133-145.

Gato, L. M. C. and Falcão, A. F. de O. 1990 Performance of Wells turbine with double row of guide vanes. *JSME International Journal, Series II*, **33** (2), 265-271.

Gato, L.M.C. and Falcão, A.F. de O. 1991 Performance of the Wells turbine with variable pitch rotor blades. *Trans ASME, J. Energy Resources Technology*, **113**, 141-146.

Gato, L.M.C., Warfield, V. and Thakker, A. 1993 Performance of a high-solidity Wells turbine for an OWC wave power plant. In *1993 European Wave Energy Symposium. Proceedings of an International Symposium held in Edinburgh, Scotland, 21-24 July 1993*, 181-189.

Gato, L.M.C., Warfield, V. and Thakker, A. 1996 Performance of a high-solidity Wells turbine for an OWC power plant. *Trans ASME, J. Energy Resources Technology*, **118** (4), 263-268.

Golomb, D. 1993 Ocean disposal of CO2: Feasibility, economics and effects. *Energy Conservation Management*, **34** (9-11), 967-976.

Graw, K.-U. 1993 Scale 1:10 wave flume experiments on IIT oscillating water column wave energy device. In *Proceedings of International Symposium on Ocean Energy Development, August 1993, Muroran, Hokkaido, Japan* (ed. H. Kondo), 221-226. (ISBN 4-906457-01-0).

Graw, K.-U. 1994 Shore protection and electricity by submerged plate wave energy converter. In *1993 European Wave Energy Symposium. Proceedings of an International Symposium held in Edinburgh, Scotland, 21-24 July 1993*, 379-384. (ISBN 0-903640-84-8).

Graw, K.-U., Schimmels, S. and Lengricht, J. 2001 Quantifying the losses around the lip of an OWC by use of particle image velocimetry (PIV). In *Proc. 4th European Wave Power Conf., Aalborg, Denmark, 4-6 December, 2000* (eds. I. Østergaard & S. Iversen), 243-250. Tåstrup, Denmark: Danish Technological Institute. (ISBN 87-90074-09-2).

Green, S. 2000 New wave device developed in Australia. *CADDET Renewable Energy Newsletter*, Jun., 2000, *19-20*.

Grue, J., Mo, A. and Palm, E.1986 The forces on an oscillating foil moving near a free surface in a wave field. *Applied Mathematics*, **3**.

Grue, J., Mo, A. and Palm, E. 1988 Propulsion of a foil moving in water waves. *J. Fluid Mech.*, **186**, 393-417.

Guangzhou Institute of Energy Conversion. 2000 <http://www.giec.ac.cn/english/>

Hagerman, G. 1992 Wave resource assessment review and evaluation. In *Comprehensive Review and Analysis of Hawaii's Renewable Energy Assessments*. R. Lynette & Associates. Final Report prepared for the State of Hawaii Department of Business, Economic Development, and Tourism. Honolulu, Hawaii.

Hagerman, G. 1995a Wave power. In *Encyclopaedia of Energy Technology and the Environment* (eds A. Bisio & S.G. Boots), 2859-2907. John Wiley & Sons Inc.

Hagerman, G. 1995b A standard economic assessment methodology for renewable ocean energy projects. In *Proceedings of the International Symposium on Coastal Ocean Space Utilization (COSU'95)*, 129-138.

Hagerman, G. 1996 Wave power: An overview of recent international developments and potential U.S. projects. In *Proceedings of the 1996 Annual Conference, American Solar Energy Society, Ashville, North Carolina* (eds R. Campbell-Howe & B. Wilkins-Crowder),195-200. Boulder, Colorado: American Solar Energy Society.

Hasselmann, K., Barnett, T.P., Bouws, E., Carlson, H., Cartwright, D.E., Enke, K., Ewing, J.A., Gienapp, H., Hasselmann, D.E., Kruseman, P., Meerburg, A., Mueller, P., Olbers, D.J., Richter, K., Sell, W. and Walden, H. 1973 Measurements of wind, wave growth and swell decay during the Joint North Sea Wave Project (JONSWAP). *Erganzungsheft Deutchen Hydrograph. Zeitschrift, Reihe A(8°), N°12*, 95pp.

Hicks, D.C., Pleass, C.M. and Mitcheson, G.R. 1988 DELBOUY: Wave-powered desalination system. In *Oceans'88 Proceedings*, **3**, 1049-1055. New York, New York: Institute of Electrical and Electronics Engineers.

Hogben, N., Dacunha, N.M.C. and Ollivier, G.F. 1986 *Global Wave Statistics*. UK: Unwin Brothers.

Hong, S.W. 2001 Wave power activities in Korea. *Private Communication*: S.W. Hong, Korea Research Institute of Ships & Ocean Engineering.

Hopfe, H.H. and Grant, A.D. 1986 The wave energy module. In *Energy for Rural and Island Communities IV* (ed J. Twidell, I. Hounam & C. Lewis), 243-248. Oxford, UK: Pergamon Press.

Hotta, H. *et al.* 1996 R & D on wave power in Japan. In *Proc. 2ⁿᵈ. European Wave Power Conference, Lisbon, Portugal, 8-10 Nov., 1995* (eds. G. Elliot & K. Diamantaras), 12-13. Brussels: European Commission. (ISBN 92-827-7492-9).

Hotta, H., Miyazaki, T., Washio, Y and Ishii, S.I. 1988 On the performance of the wave power device Kaimei: The results on the open sea tests. In *Proceedings of the Seventh International Conference on Offshore Mechanics and Arctic Engineering*, 91-96. New York, New York: American Society of Mechanical Engineers.

Hotta, H., Washio, Y., Ishii, S., Masuda, Y., Miyazaki, T. and Kudo, K. 1986. The operational test on the shore fixed OWC type wave power generator. In *Proceedings of the Fifth International Offshore Mechanics and Arctic Engineering Symposium*, **2**, 546-552. New York, New York: American Society of Mechanical Engineers.

Hotta, H, Washio, Y., Yokozawa, H. and Miyazaki, T. 1996 On the open sea tests of a prototype device of a floating wave power device 'Mighty Whale'. In *Proc. 2ⁿᵈ European Wave Power Conference, Lisbon, Portugal, 8-10 Nov., 1995* (eds. G. Elliot & K. Diamantaras), 404-405. Brussels: European Commission. (ISBN 92-827-7492-9).

Institute of Energy (UK). 2001 World's first wave power device feeds power to Scotland. *Energy World*, (286), Feb. 2001, 13.

Institute of Energy (UK). 2002a Wave energy project planned for Canada. *Energy World*, (299), May 2002, 4.

Institute of Energy (UK). 2002b Boost for wave energy, and the Western Isles. *Energy World*, (301), Jul./Aug. 2002, 3.

International Desalination Association. 1998 *Report No.15*. International Desalination Association,

International Energy Agency. 1994a *Energy Prices and Taxes-1994*. Paris: IEA/OECD.

International Energy Agency. 1994b *Energy and Environmental Technologies to Respond to Global Climate Change Concerns*. Paris, France: OECD/IEA.

International Energy Agency. 1998 *Renewables in power generation: Towards a better environment*. Paris, France: OECD/IEA.

International Patent Application. 1995 No. WO 95/17555.

Interproject Services AB. 2000 <http://www.members.tripod.com/interproject>

Ippen, A.T. (ed.). 1966 *Estuary and Coastline Hydrodynamics*. New York, NY: McGraw-Hill.

Jakobsen, E. 1981 The foil propeller: wave power for propulsion. In *Proc. 2ⁿᵈ Int. Symposium on Wave and Tidal Energy, Cambridge, UK, 23-25 Sep., 1981*, 363-388. Cranford, UK: BHRA Fluid Engineering. (ISBN 0-906085-43-9).

Justino, P.A.P., Nichols, N.K. and Falcão, A.F. de O 1994 Optimal phase control of OWCs. In *1993 European Wave Energy Symposium. Proceedings of an International Symposium held in Edinburgh, Scotland, 21-24 July 1993*, 145-149.

Justino, P. A. P. and Falcão, A. F. de O. 1999 Rotational speed control of an OWC wave power plant. *Trans. ASME - Journal of Offshore Mechanics and Arctic Engineering*, 121 (4), 65-70.

Kayser, H. 1974 Energy generation from sea waves. In *Proceedings Ocean '74 IEEE Conference, Halifax, N.S., Canada*, 1, 240-243.

Kingston, W., Lindblom, U. and Bergdahl, L. 2000 The Rock OWC. In *Proc. 3ʳᵈ. European Wave Power Conf., Patras, Greece, 30 Sep.-2 Oct., 1998* (ed. W. Dursthoff). University of Hannover.

Kofoed, J.P., Frigaard, P., Sorenson, H.C. and Friis-Madsen, E. 2000 Development of the wave energy converter - Wave Dragon. In *Proc. 10th (2000), ISOPE Conference, Seattle, USA, 28 May-2 Jun., 2000*.

Komen, G., Cavaleri, L., Donelan, M., Hasselmann, K., Hasselmann, S. and Janssen, P.A.E.M. 1994 *Dynamics and Modelling of Ocean Waves*. Cambridge Univ. Press.

Kondo H 1998 The impact on environment and marine industries. *Unpublished manuscript*.

Korde, U.A. 2002 Active control in wave energy conversion. *Sea Technology*, 43 (7), 47-52.

Kyllingstad, Å. 1982 Approximate analysis concerning wave-power absorption by hydrodynamically interacting buoys. *Thesis for the "dr.ing." degree*, University of Trondheim, Norwegian Institute of Technology, Division of Experimental Physics.

Laboratorio Nacional de Engenharia e Tecnologia Industrial, Portugal. 1993 Preliminary actions in wave energy R & D, wave studies and development of resource evaluation methodology. *Technical Report EU-DG XII Contract No JOUR-0132-PT (MNRE)*, 75p. The European Economic Communities.

Liang, X. and Wang, W. 1996 Experimental research on the performance of a BBDE wave-activated generation device model. In *Proc. 2ⁿᵈ. European Wave Power Conf., Lisbon, Portugal, 8-10 Nov., 1995* (eds. G. Elliot & K. Diamantaras). Brussels: European Commission. (ISBN 92-827-7492-9).

Lillebekken, P.M., Aakenes, U.R. and Falnes, J. 2000 The ConWEC wave energy device. In Proc. 3ʳᵈ European Wave Power Conf., Patras, Greece, 30 Sep.-3 Oct., 1998 (ed. W. Dursthoff). University of Hannover.

References cited

Lockett, F.P. 1991 The Clam wave energy converter. In *Proceedings of a Seminar on Wave Energy*. London, UK: Institute of Mechanical Engineers.

Maisondieu, C. 1993 Absorption dynamique des ondes de gravité en régime instationnaire. Thèse d'Université, ENSM, Nantes.

Malmo, O. and Reitan, A. 1986. Development of the Kvaerner multiresonant OWC. In *Proc. IUTAM Symposium on Hydrodynamics of Ocean Wave-energy Utilization, Lisbon, Portugal, 8-11 July, 1985* (eds. D.V. Evans & A.F. de Falcao), 57-67. Springer-Verlag.

Marton, I. 1991 Bølgekraftverk i fjell - Bølgeorgel (Wave power plant in rock - wave organ). *SINTEF Report STF 60 A91070*. Trondheim, Norway: NHL.

Masuda, Y., Yamazaki, T., Outa, Y. and McCormick, M.E. 1987 Study of backward bent duct buoy. In *Oceans Proceedings*, **2**, 384-389. Washington, DC: Marine Technology Society.

Mavrakos, S.A. and Kalofonos, A. 1996 Optimum power absorption by arrays of interacting vertical axisymmetric wave-energy devices. In *Proc. 15th. OMAE, Florence, 1996*, **1** (Part B).

Mavrakos, S.A. and Kalofonos, A. 1997 Optimum power absorption by arrays of interacting vertical axisymmetric wave-energy devices. *Journal of Offshore Mechanics and ArcticEngineering.* **119**, 244 - 251.

Mavrakos, S.A. and McIver, P. 1997 Comparison of methods for computing hydrodynamic characteristics of arrays of wave power devices. *Applied Ocean Research*, **19**, 283-291.

McAlpine, Sir Robert and Sons Ltd. 1982 Final report on the optimisation of the Bristol submerged cylinder wave energy device. *ETSU Report WV-1630*. Harwell, UK: Department of Trade & Industry.

McCormick, M.E. 1974 Analysis of a wave energy conversion buoy. *Journal of Hydronautics*, **8** (3), 77-82.

McCormick, M.E. 2001 Wave-powered reverse-osmosis desalination. *Sea Technology*, **42** (12), 37-39.

McCormick, M.E., McCabe, R.P., and Kraemer, D.R.B. 1998 Potable water and electricity production by a hinged-barge wave energy conversion system. *Int. J. Power & Energy Systems.*

Mehlum, E. 1982 Recent developments in the focusing of wave energy. In *Proc. 2nd Int. Symposium on Wave Energy Utilisation* (ed H. Berge), 419-420. Trondheim, Norway: Tapir. (ISBN 82-519-0478-1).

Mehlum, E. 1986 TAPCHAN. *Proc. IUTAM Symposium on Hydrodynamics of Ocean Wave-energy Utilization, Lisbon, Portugal, 8-11 July, 1985* (eds. D.V. Evans & A.F. de Falcao), 51-55. Springer-Verlag.

145

Mei, C.C. 1976 Power extraction from water waves. *Journal of Ship Research*, **21** (4), 248-253.

Milgram, J.H.. 1970 Active water-wave absorbers. *Journal of Fluid Mechanics*, **43**, 845-859.

Miyazaki, T. 1991 Wave energy research and development in Japan. In *Oceans '91, Honolulu, Hawaii*. Alexandria, Virginia: SEASUN Power Systems (reprint).

Miyazaki, T., Washio, Y. and Kato, N. 1993 Performance of the floating wave energy converter Mighty Whale. In *Proceedings of International Symposium on Ocean Energy Development for Overcoming Energy & Environmental Crises*, 197-204. Muroran, Hokkaido, Japan: Muroran Institute of Technology.

Moreira, N.M., Pires, H.O., Pontes, M.T. and Camara, C. 2002 Verification of TOPEX/Poseidon wave data against buoys off the west coast of Portugal. In *Proc. 21ˢᵗ International Conf. on Offshore Mechanics and Arctic Engineering, Oslo, Norway, Jun., 2001*, Paper No. 28254. OMAE.

Naito, S. and Nakamura, S. 1986 Wave energy absorption in irregular waves by feed-forward control system. In *Hydrodynamics of Ocean Wave-Energy Utilization, IUTAM Symposium, Lisbon, 1985*, (eds. D.V. Evans & A.F. de O. Falcão), 269-280. Berlin: Springer Verlag.

National Institute of Ocean Technology. 2000 <http://www.niot.ernet.in>

NAVFACCO. 1997. Natural Resource Development at NSF Diego Garcia (Advertisement). Sea Technology, **38** (4), 148.

Nebel, P. 1992 Maximising the efficiency of wave energy plants using complex-conjugate control. *Proc. I. Mech. E., Journal of Systems & Control*. **206** (Part 1), 225-236.

Newman, J.N. 1976 The interaction of stationary vessels with regular waves. In *Proc. 11th Symp. on Naval Hydrodynamics, London, UK*, 491-501.

Nielsen, K., Scholten, N.C. and Sørensen, K.A . 1996 Hanstholm offshore wave energy experiment. In *Proc. 2nd. European Wave Power Conference, Lisbon, Portugal, 8-10 Nov., 1995* (eds. G. Elliot & K. Diamantaras), 44-51. Brussels: European Commission. (ISBN 92-827-7492-9).

Nielsen, K. and Meyer, N.I. 2000 The Danish wave energy programme. In *Proc. 3rd. European Wave Power Conf., Patras, Greece, 30 Sep.-2 Oct., 1998* (ed. W.Dursthoff). University of Hannover.

Nielsen, K. and Scholten, C.1990 Planning a full scale wave power conversion test:1988/1989. In *Proc. International Conference on Ocean Energy Recovery* (ed. Hans-Jurgen Krock), 111-120. New York, New York: American Society of Civil Engineers.

Nielsen, K and Smed, P.F. 2000 Point absorber--optimisation and survival testing. In *Proc. 3ʳᵈ European Wave Power Conf., Patras, Greece, 30 Sep.-2 Oct., 1998* (ed. W.Dursthoff). University of Hannover.

146

References cited

Norse, E.A. (ed). 1993 *Global Marine Biodiversity: A Strategy for Building Conservation into Decision Making*, 106-115. Washington, DC: Island Press.

Norwegian Royal Ministry of Petroleum and Energy. 1987 *Norwegian wave power plants 1987*. Oslo, Norway: Royal Ministry of Petroleum and Energy.

Ocean Power Delivery. 2000 <http://www.oceanpd.com>

OCEANOR. 2002 <http://www.oceanor.no/projects/wave_energy>

Ohno, M., Funakoshi, H., Saito, T., Oikawa, K. and Takahashi, S. 1993 Interim report on the second stage of field experiments on a wave power extracting caisson in Sakata Port. In *Proceedings of International Symposium on Ocean Energy Development for Overcoming the Energy & Environmental Crises*,173-182. Muroran, Hokkaido, Japan: Muroran Institute of Technology.

Olje-og energidepartementet 1982. *Nye fornybare energikilder i Norge*. (*New renewable energy sources in Norway*.) Stortingsmelding nr. 65, 1981-82 (White paper No. 65, 1981-82). Oslo, Norway: Royal Ministry of Petroleum and Energy.

Oliveira-Pires, H.,Carvalho, F. and Pontes, T. 1996 Modeling the effect of shelter in the modification of waves from the open sea to near-shore. *Transactions of ASME - Journal of Offshore Mechanics and Arctic Engineering*, **119**, 70-72.

Oliveira-Pires, H., Pontes. M.T. and Aguiar, R. 1999 Directional statistics and the characterization of the wave energy resource. In *Proc. 18th International Conference on Offshore Mechanics and Arctic Engineering - OMAE'99, July, St. Johns´s, Newfoundland, Canada*, (Paper #3112).

Osanai, S., Narita, M., Kondo, H., Mizuno, Y. and Watabe, T. 1996 1995 feasibility tests of new pendular-type wave energy conversion apparatus. In *Proc. 2nd. European Wave Power Conference, Lisbon, Portugal* (eds. G. Elliot & K. Diamantaras), 320-323. Brussels: European Commission. (ISBN 92-827-7492-9).

Panicker, N.N. 1976 Power resource potential of ocean surface waves. In *Proceedings of the Wave and Salinity Gradient Workshop, Newark, Delaware, USA*, (Paper J1-J48).

Peatfield, T. 1991 The economic viability of the circular clam for offshore wave energyutilisation. In *Proceedings of a Seminar on Wave Energy*. London, UK: Institute of Mechanical Engineers.

Perdigão, J. N. A. 1998 Reactive-Control Strategies for an Oscillating-Water-Column Device. *Ph. D. Thesis*, Lisbon Technical University.

Perdigão, J.N.B.A. and Sarmento, A.J.N.A. 1989 A phase control strategy for OWC devices in irregular seas. In *Proc.The Fourth International Workshop on Water* (ed. J. Grue).

147

Pierson, W. and Moskowitz, L. 1964 A proposed spectral form of fully developed wind seas based on the similarity theory of S. A. Kitaigorodskii. *J. Geophysical Res.* **69**, 5181-5190.

Pizer, D.J. 1993 Maximum wave power absorption of point absorbers under motion constraints. Applied Ocean Research, **15**, 227-234.

Platts, M.J. 1982 Engineering design of the Lancaster wave energy system. In *Proceedings of the Second International Conference on Wave Energy Utilization* (ed H. Berge), 253-274. Trondheim, Norway: Tapir Publishers.

Pontes, M.T. 1998 Assessing the European wave energy resource. *Trans. ASME - Journal of Offshore Mechanics & Offshore Engineering*, **120**, 226-231.

Pontes, M.T., Athanassoulis, G.A., Barstow, S., Cavaleri, L., Holmes, B., Mollison, D. and Oliveira-Pires, H. 1996a Atlas of wave energy resource in Europe. *Technical* Report, EU *DGXII, Contract N°JOU2-CT93-0390.*

Pontes, M.T., Aguiar, R.and Oliveira-Pires, H. 2001 A near-shore wave energy atlas for Portugal. In *Proc. 4th European Wave Power Conf., Aalborg, Denmark, 4-6 December, 2000* (eds. I. Østergaard & S. Iversen), 72-9. Tåstrup, Denmark: Danish Technological Institute (ISBN 87-90074-09-2).

Pontes, M.T., Bertotti, L., Cavaleri, L. and Oliveira-Pires, H. 1996b Use of numerical wind-wave models for wave energy resource assessment. *Trans. ASME -Journal of Offshore Mechanics & Arctic Engineering.*, **119**, 184-190.

Pontes, M.T., Mollison, D., Cavaleri, L., Nieto, J.C. and Athanassoulis, G.A. 1993 Wave studies and development of resource evaluation methodology. *Final Report*, *EU DGXII, Contract N° JOUR-0132-PT(MNRE)*, 75p.

POSEIDON. 2001 <http://www.poseidon.ncmr.gr>

Raghunathan, S.R. 1980 The Wells turbine. *The Queen's University of Belfast, Department of Civil Engineering, Report No. WE/80/11.* Belfast, UK: Queen's Univ. of Belfast.

Rampen, W.H.S., Almond, J.P., Taylor, J.R.M., Eshan, M. and Salter, S.H. 1996 Progress on the development of the wedding-cake digital hydraulic pump/motor. In *Proc. 2nd. European Wave Power Conf., Lisbon, Portugal, 8-10 Nov., 1995* (eds. G. Elliot & K. Diamantaras), 289-296. Brussels: European Commission. (ISBN 92-827-7492-9).

Ravindran, M., Jayashankar, V., Jalihal, P. and Pathak, A.G. 1997 The Indian wave energy programme - - an overview. *Tide (TERI Information Digest on Energy)*, **7** (3),173-188.

Rebollo, L., Matas, A., Wilhelmi, J.R., Fraile, J.J., Laguna, F.V., Martinez, M. and Berenguer, J.M. 1996 Project OLAS 1000 – experiences, modelling and results. In *Proc. 2nd. European Wave Power Conference, Lisbon, Portugal, 8-10 Nov., 1995* (eds. G. Elliot & K. Diamantaras), 312-319. Brussels: European Commission. (ISBN 92-827-7492-9)..

References cited

Retzler, C.H. 1996 The wave rotor. In *Proc. 2nd. European Wave Power Conf., Lisbon, Portugal, 8-10 Nov., 1995* (eds. G. Elliot & K. Diamantaras), 329-336. Brussels: European Commission. (ISBN 92-827-7492-9).

Royal Commission on Environmental Pollution. 2000 *Energy–the changing climate*. London, UK: HMSO.

Russel, A. and Diamantaras, K. 1996 The European Commission Wave Energy R & D Programme. In *Proc. 2nd. European Wave Power Conf., Lisbon, Portugal, 8-10 Nov., 1995* (eds. G. Elliot & K. Diamantaras), 8-11. Brussels: European Commission. (ISBN 92-827-7492-9).

SABRE-Gen. 2002 <http://www.sabregen.co.za>

Salter, S.H. 1974 Wave power. *Nature*, **249**, 720-724.

Salter, S.H. 1979 Power conversion systems for ducks. In *Proc. Int. Conf. on Future Energy Concepts, January 1979, London, UK* (Publication 171), 100-108. London, UK: Institution of Electrical Engineers,

Salter, S.H. 1985 Progress on Edinburgh Ducks. In *Proceedings of IUTAM Symposium on Ocean Wave Energy Utilisation, Lisbon, Portugal*, 36-50.

Salter, S.H.. 1989 World progress in wave energy--1988. *International Journal of Ambient Energy*, **10**, 3-24.

Salter, S.H. 1994a Variable pitch air turbines. In *1993 European Wave Energy Symposium. Proceedings of an International Symposium held in Edinburgh, Scotland, 21-24 July 1993*, 435-442. (ISBN 0-903640-84-8).

Salter, S.H. 1994b Changes to the 1981 design of spine-based Ducks. In *1993 European Wave Energy Symposium. Proceedings of an International Symposium held in Edinburgh, Scotland, 21-24 July 1993*, 259-309. (ISBN 0-903640-84-8).

Salter, S.H., Jeffery, D.C. and Taylor J.R.M. 1976 The architecture of nodding duck wave power generators. *The Naval Architect*, Jan 1976, 21-24.

Salter, S.H. and Lin, C-P. 1996 The sloped IPS wave energy converter. In *Proc. 2nd. European Wave Power Conference, Lisbon, Portugal, 8-10 November, 1995* (eds. G. Elliot & K. Diamantaras). Brussels: European Commission. (ISBN 92-827-7492-9).

Salter, S.H. and Rampen, W.H.S. 1993 The wedding cake multi-eccentric radial piston hydraulic machine with direct computer control of displacement. In *Proc. 10th. International Conference on Fluid Power, Brugge, The Netherlands*, 47-64.

Salter, S.H. and Taylor, J. 1996 The design of a high-speed stop valve for oscillating water columns. *Proc. 2nd. European Wave Power Conference, Lisbon, Portugal, 8-10 November, 1995* (eds.

149

G. Elliot & K. Diamantaras), 195-202. Brussels: European Commission. (ISBN 92-827-7492-9).

Sarmento, A.J.N.A. 1993 Model tests optimisation of an OWC wave power plant. *Int. J. Offshore and Polar Engng.*, **3** (1), 66-72.

Sarmento, A.J.N.A., Gato, L.M.C. and Falcão, A.F. de O. 1990 Turbine-controlled wave energy absorption by oscillating water column devices. *Ocean Engineering*, **5**, 481-497.

Santhakumar, S., Jayashankar, V., Atmanand, M.A., Pathak, A.G., Ravindran, M., Setoguchi, T., Takao, M. and Kaneko, K. 1998 Performance of an impulse turbine based wave energy plant. In *Proceedings of the Eighth International Offshore and Polar Engineering Conference, Montreal, Canada, May, 1998*, **1**, 75-80. The International Society of Offshore and Polar Engineers.

Schleisner, L.and Nielsen, P. 1997 Environmental external effects from wind power based on the EU ExternE methodology. Paper presented at EWEC 1997, the European Wind Energy Conference and Exhibition, Dublin.

Scott, N.C. 2001 Grid connection of large-scale wave energy projects: an electrical overview. In *Proc. 4th European Wave Power Conf., Aalborg, Denmark, 4-6 December, 2000* (eds. I. Østergaard & S. Iversen), 88-92. Tåstrup, Denmark: Danish Technological Institute . (ISBN 87-90074-09-2).

Sea Energy Associates. 1986 The development of the circular SEA Clam. *ETSU Report No. WV1676*, Sea Energy Associates and Lancaster Polytechnic, UK.

SEASUN Power Systems. 1988 Power conditioning and transmission, wave energy resource and technology assessment for coastal North Carolina. *Final Report, 5.1-5.21.* North Carolina Alternative Energy Corporation & Virginia Power,

SEASUN Power Systems. 1992 Wave energy resource and economic assessment for the State of Hawaii. Honolulu, Hawaii: Department of Business and Economic Development, Energy Division.

Setoguchi, T., Kaneko., K., Maeda, H. and Raghunathan, S. 1994 Impulse turbine with self-pitch-controlled guide vanes for wave energy conversion. In *1993 European Wave Energy Symposium. Proceedings of an International Symposium held in Edinburgh, Scotland, 21-24 July 1993*, 203-208. (ISBN 0-903640-84-8).

Shackell, N.L. and Willison, J.H.M. 1995 Marine Protected Areas and Sustainable Fisheries. *In Proc. of a symposium on marine protected areas and sustainable fisheries conducted at the Second Int. Conf. on Science & Management of Protected Areas, Halifax, Nova Scotia, Canada, 16-20 May, 1994*, 300p. Wolfville, NS, Canada: Science and Management of Protected Areas, Acadia Univ.

Shiki, A., and Iwase K. 1990 Research and development of wave pump. In *Proceedings of the International Conference on Ocean Energy Recovery* (ed Hans-Jurgen Krock), 76-83. New York, New York: American Society of Civil Engineers.

Shim, H.J. 1996 Electric generator using wave forces. *Private communication*: H.J. Shim, Baek Jae Engineering, Seoul, Korea.

Side, J.C. 1992 *Decommissioning and abandonment of offshore installations, North Sea oil and the environment; developing oil and gas resources, environmental impacts and responses* (ed W.J. Cairns), 523-545. Elsevier Applied Science.

Sjöström, B-O. 1994 The past, present and future of the hose-pump wave energy converter. In *1993 European Wave Energy Symposium. Proceedings of an International Symposium held in Edinburgh, Scotland, 21-24 July 1993*, 311-316. (ISBN 0-903640-84-8).

Sjöström, B-O., Economides, J. and Gallant, G. 1996 The Amorgos Project, using wave energy for fresh water and electricity production. In *Proc. Mediterranean Conference on Renewable Energy Sources for Water Production*, Santorini, 217-221. (ISBN 960-90557-0-2).

Southgate, H.N.1993 The use of wave transformation models to evaluate inshore wave energy resources. In *1993 European Wave Energy Symposium. Proceedings of an International Symposium held in Edinburgh, Scotland, 21-24 July 1993*, 41-46. (ISBN 0-903640-84-8).

Spiridakis, M.1983 Récupération de l'énergie des vagues par les systèmes à déferlement. *Thèse d'Université*. Nantes, France: ENSM.

Suroso, A. 2001 Wave power activities in Indonesia. *Private Communication*: A. Suroso, Sepuluh Nopember Institute of Technology, Surabaja, Indonesia.

Takahashi, S. 1988 *A study on the design of a wave power extracting caisson breakwater*. Japan: Port and Harbour Research Institute.

Tanaka, Y., Furukawa, K. and Motora, Y. 1993 Experimental and theoretical study of a double OWC floating wave power extractor. In *Proc. International Symposium on Ocean Energy Development for Overcoming the Energy & Environmental Crises*, 209-220. Muroran, Hokkaido, Japan: Muroran Institute of Technology.

Taylor, J.R.M. and Salter, S.H. 1996 Design and testing of a plano-convex bearing for a variable-pitch turbine. In *Proc. 2nd. European Wave Power Conference, Lisbon, Portugal, 8-10 Nov., 1995* (eds. G. Elliot & K. Diamantaras), 240-247. Brussels: European Commission. (ISBN 92-827-7492-9).

Taylor, G. 1999 OPT Wave power system. In *Wave Power, Moving Towards Commercial Viability, International One Day Seminar, 30 Nov., 1999*. London, U.K.: Institution of Mechanical Engineers.

Thomas, G.P. and Evans, D.V. 1981 Arrays of three-dimensional wave-energy absorbers. *Journal of Fluid Mechanics*, **108**, 67-88.

Thorpe, T.W. 1992 A review of wave energy. *ETSU Report, R-72*, **1, 2**. Harwell, UK: Department of Trade & Industry.

Thorpe, T.W. 1995 An assessment of the Art Osprey wave energy device. *ETSU Report, R-90*. Harwell, UK: Department of Trade and Industry.

Thorpe, T.W. 1997 Assessment of the Shim wind-wave energy device. *ETSU Report, AEAT-1391*. Harwell, UK: Department of Trade and Industry.

Thorpe, T.W. 1998 An overview of wave energy technologies for the UK Marine Foresight Panel. *ETSU Report, AEAT-3615*. Harwell, UK: Department of Trade and Industry.

Thorpe, T.W. 1999a A brief overview of wave energy. *ETSU Report, R-120*. Harwell, UK: Department of Trade and Industry.

Thorpe, T.W. 1999b An overview of wave energy technologies: status, performance and costs. In *Wave Power, Moving Towards Commercial Viability, International One Day Seminar, 30 Nov., 1999*. London, U.K.: Institution of Mechanical Engineers.

Thorpe, T.W. 2000 An evaluation of wave energy. *ETSU Report AEAT/ENV/R/0400*. Oxfordshire, U.K.: Department of Trade and Industry.

Thorpe, T.W. and Picken, M.J. 1993 Wave energy devices and the marine environment. *IEE Proceedings*. **A140** (1), 63-70.

Tjugen, K.J. 1996 TAPCHAN Ocean Wave Energy Project at Java: Updated Project Status. In *Proc. 2nd. European Wave Power Conference, Lisbon, Portugal, 8-10 November, 1995* (cds. G. Elliot & K. Diamantaras), 42-43. Brussels: European Commission. (ISBN 92-827-7492-9).

Tveter, T. 2001 Wave-pump-energy. In *Proc. 4th European Wave Power Conf., Aalborg, Denmark, 4-6 December, 2000* (eds. I. Østergaard & S. Iversen), 93-97. Tåstrup, Denmark: Danish Technological Institute (ISBN 87-90074-09-2).

UK Government. 2000 <http://www.parliament.the-stationary-office.co.uk/pa/cm200001/cmselect/cmsctech/291/29102.htm>

University of Edinburgh, Wave Power Group. 2000 <http://www.mech.ed.ac.uk/research/wavepower/>

U.S. Patent. 1988 U.S. Patent 4, 781, 023 (to C.K. Gordon).

van Zanten, W. 1996 Archimedes Wave Swing. *CADDET Newsletter*, (4/96).

Various authors. 1993 Preliminary design and model test of a wave-power converter: Budal's 1978 design Type E. *Technical reports compiled by J. Falnes, Department of Physics, Norwegian*

University of Science and Technology. Tronheim, Norway: Norwegian University of Science and Technology

Vogel, S. 1981 *Life in Moving Fluids: The Physical Biology of Flow*. Princeton, NJ: Princeton Univ. Press, 352 p.

Walton-Bott, A.N., Hunter, P.D. and Hailey, J.S.M. 1988 The Mauritius wave energy scheme - present status. In *Euromechanics Colloquium 243, Energy from Ocean Waves* (extended abstract), (ed D.V. Evans). Bristol, UK: University of Bristol.

WAMDI Group. 1988 The WAM model - a third generation ocean wave prediction model. *J. Phys. Oc.*, **18**, 1775-1810.

Wangwick, K. 1998 55% Higher sales and dramatic changes in market. *Desalination & Water Reuse*, **8** (2), 11-13.

Watabe, T. *et al.* 1996 Rotary vane pump. *Japanese Patent*, 2573905.

Watabe, T. and Kondo, H. 1989 Hydraulic technology and utilization of ocean wave power. In *JHPS International Symposium on Fluid Power, Tokyo, Japan, March 1989*, 301-308.

Water Environment Transport, Chalmers University of Technology. 2000 <http://www.wet.chalmers.se>

Wavegen. 2000 <http://www.wavegen.co.uk>

Whittaker, T.J. and McIlwaine, S. 1991 Shoreline wave power experience with the Islay prototype. In *Proceedings of the First (1991) International and Polar Engineering Conference*, **1**, 393-397. International Society of Offshore and Polar Engineers.

Whittaker, T.J. and Raghunathan, S. 1993 A review of the Islay shoreline wave power station. In *1993 European Wave Energy Symposium. Proceedings of an International Symposium held in Edinburgh, Scotland, 21-24 July 1993*, 283-294 (ISBN 0-903640-84-8).

Whittaker, T J. *et al.* 1991 Islay shoreline wave energy device – Phase 2. *ETSU Report No. WV1680*. Harwell, U.K.: Department of Trade & Industry

Whittaker, T J. *et al.* 1996a Implications of operational experience of the Islay OWC for the design of Wells turbines. In *Proc. 2nd. European Wave Power Conf., Lisbon, Portugal, 8-10 November, 1995* (eds. G. Elliot & K. Diamantaras). Brussels: European Commission. (ISBN 92-827-7492-9).

Whittaker, T J. *et al.* 1996b Islay European shoreline wave power plant, the design of a modular system. In *Proc. 2nd. European Wave Power Conference, Lisbon, Portugal, 8-10 November, 1995* (eds. G. Elliot & K. Diamantaras). Brussels: European Commission. (ISBN 92-827-7492-9).

Wilke, R.O. 1989 First open-water test program for a full-scale model of the tandem flap wave energy conversion device. In *Proceedings of the Eighth International Conference on Offshore Mechanics and Arctic Engineering*, 435-438. New York, New York: American Society of Mechanical Engineers.

World Energy Congress. 1989 The potential of renewable energy, an inter-laboratory paper. In *14th Congress of the World Energy Council, Montreal, Canada, September 1989.*

World Energy Congress. 1993 Renewable energy resources: Opportunities and constraints 1990-2020. In *Proc. World Energy Council, London, September 1993.*

World Meteorological Organization. 1976 *Handbook on Wave Analysis and Forecasting* (WMO No.446). World Meteorological Organization.

Wu, F.H.Y. and Liao, T.T.L. 1990 Wave power development in Taiwan. In *Proceedings of the International Conference on Ocean Energy Recovery* (ed H-J. Krock), 93-100. New York, New York: American Society of Civil Engineers.

Yazaki, A.., Takezawa, S. and Sugawara, K. 1986 Experiences on field tests of "Kaiyo" electricity generating system by natural wave energy. In *Proceedings of the Fifth International Offshore Mechanics and Arctic Engineering Symposium*, **2**, 538-545. New York, New York: American Society of Mechanical Engineers.

Yeaple, F. 1989 Wave-driven reverse osmosis makes fresh water cheaply. *Design News*, March 1989.

Young, I.R. and Holland, G.J. 1996 *Atlas of the Oceans: Wind and Wave Climate*. Pergamon. (ISBN 0-08-042519-4).

Yu, Z. 1995 Development and utilization of ocean wave energy in China. In *Proceedings of World Solar Summit Process, Solar Energy in China, Beijing, China*, 192-199.

Yu, Z. and You, Y. 1996 Modelling and analysis of a prototype onshore wave power station. In *Proc. 2nd. European Wave Power Conf., Lisbon, Portugal, 8-10 Nov., 1995* (eds. G. Elliot & K. Diamantaras). Brussels: European Commission. (ISBN 92-827-7492-9).

Appendix 1

WORKING GROUP MEMBERS
(SC indicates Steering Committee Member)

Mr John Brooke, Vice-President, ECOR (SC) (Chair) <az337@chebucto.ns.ca>	*Summer address*: R.R.#1, Pleasantville, NS, B0R 1G0, Canada. *Winter address*: Suite 409, Admiralty Place, 1 Prince St., Dartmouth, NS, B2Y 4L3, Canada	T: (902) 543-2425 T: (902) 461-4151
Dr. Teresa Pontes (SC) (Vice-Chair) <teresa.pontes@ineti.pt>	INETI – Department of Renewable Energies, Estrada do Paco do Lumiar, 1649-038, Lisbon, Portugal	T: +351 21 7127201 F: +351 21 7127195
Prof. Lars Bergdahl (SC) <lars.bergdahl@wet.chalmers.se>	Water Environment Transport, Chalmers University of Technology, SE-412 96, Gothenburg, Sweden	T: +46 31 772 21 55 F: +46 31 772 21 28
Prof. Johannes Falnes (SC) <falnes@phys.ntnu.no>	Department of Physics, Norwegian Univ. of Science & Technology, N-7034, Trondheim, Norway	T: +47-735-93452 F: +47-735-97710
Prof. Kenji Hotta (SC) <hotta@ocean.cst.nihon-u.ac.jp>	Department of Oceanic Architecture & Engineering, Nihon University, 7-24-1 Narashinodai Funabashi-Shi Chiba, 274850 Japan	T: +81 474 69 5458 F: +81 474 67 9446
Mr. Thomas Thorpe (SC) <tom.thorpe@tiscali.co.uk>	Energetech Australia Pty 1 The Avenue Randwick 2031, Australia	M: +44 7787 534770 T: +61 2 9326 4237 F: +61 2 9326 6277

Prof. Neil Bose <nbose@engr.mun.ca>	Memorial University of Newfoundland, Faculty of Engineering. & Applied Science, St. John's, NF, A1B 3X5, Canada	T: (709) 737-4058 F: (709) 737-2116
Mr. George Hagerman <hagerman@vt.edu>	Center for Energy & the Global Environment, Virginia Tech Alexandria Research Institute, 206 N. Washington St. (# 400), Alexandria, VA 22314, USA	T: (703) 535-3461 F: (703) 518-8085
Dr. Hideo Kondo <cosinehk@smile.ocn.ne.jp>	Coastsphere Systems Institute, #1003 TM Bldg., 15-1 N7 W2 Kita-Ku, Sapporo, 060-0807, Japan	T: +81-11-738-3320 F: +81-11-738-3321
Prof. Michael McCormick	U.S. Naval Academy, Mail Stop 11d, 590 Holloway Rd., Annapolis, MD 21402-5042, USA	
Prof. M. Ravindran <mravin@niot.ernet.in>	National Institute of Ocean Technology, IC & SR Building, IIT Campus, Chennai-600 036, India	T: +91 44 23 53684 F: +91 44 23 3686/2545
Prof. Zhi YU <yuzhi@ms.giec.ac.cn>	Dept. of Applied Mechanics & Engineering, Zhong Shan University, CAS Guangzhou, 510070, P.R.China	T: +86 20 87606993 F: +86 20 87302770

Reporting Editor: Mr. Brian Nicholls <nicholls@telusplanet.net>	Suite 1201, 9921-104 St., Edmonton, AB, T5K 2K3, Canada	T: (780) 426-3278 F: (780) 426-5097

Appendix 2

MATHEMATICAL DESCRIPTION OF WAVES AND WAVE ENERGY[1]

Hydrodynamics of Sea Waves

The creation of waves is a complex, nonlinear process in which energy is slowly exchanged between different components (see Komen *et al.* 1994). However, on a scale of tens of kilometres and minutes in deep water, a stationary Gaussian random process accurately describes the local state of the sea surface. Thus the local behaviour of the waves is determined by the spectrum of the sea state $S(f,\theta)$ that specifies how the wave energy, proportional to the variance of the surface elevation, is distributed in terms of frequency f and direction θ. This spectrum can in turn be summarised by a small number of wave parameters, namely wave height H, period T $(f=1/T)$, and direction.

For wave height, the most widely used parameter is the significant wave height, defined as the average of the highest one third of the trough to crest wave heights, and matching reasonably well one's visual impression of wave height. It can be computed from the spectrum by

$$H_s = 4m_0^{1/2}$$
(1)

where m_0 is the zero-th spectral moment, the n-th moment being defined as

$$m_n = \int_0^{2\pi} \int_0^{\infty} f^n S(f,\theta)\, df\, d\theta$$
(2)

For wave period, several parameters are commonly used. The most appropriate, for present purposes, are the mean (energy) period T_e and the peak period T_p. The energy period is defined by

$$T_e = \frac{m_{-1}}{m_0}$$
(3)

Since $T=1/f$, T_e is the average value of T over the wave spectrum, T_e depends mainly on the lower frequency band of the spectrum where most of the energy is contained. It is thus more stable than the traditional zero-crossing period T_{m02}, the average time elapsed between two sequential crests computed by $(m_0 / m_2)^{1/2}$. Its dependence on m_2 makes it very sensitive to the high frequency spectral tail that exhibits high variability and minute energy contents.

The peak period T_p is the inverse of the peak frequency f_p that corresponds to the highest spectral density

$$T_p = \frac{1}{f_p}$$

(4)

Several wave direction parameters can be used. Taking the directional spectrum, mean wave direction is computed by

$$\bar{\theta} = \arctan\left(\frac{\int_0^{2\pi}\int_0^\infty S(f,\theta)\sin(\theta)\mathrm{d}\theta\,\mathrm{d}f}{\int_0^{2\pi}\int_0^\infty S(f,\theta)\cos(\theta)\mathrm{d}\theta\,\mathrm{d}f}\right).$$

(5)

Directional buoys have often provided only frequency spectra $E(f)$, related to the directional spectrum by

$$E(f) = \int_0^{2\pi} S(f,\theta)\mathrm{d}\theta$$

(6)

in addition to the mean direction $\theta(f)$ and its spreading for each frequency. Mean direction is then computed by

$$\bar{\theta}_b = \arctan\left(\frac{\int_0^\infty E(f)\overline{\sin(\theta(f))}\,\mathrm{d}f}{\int_0^\infty E(f)\overline{\cos(\theta(f))}\,\mathrm{d}f}\right).$$

(7)

Often, an oceanic sea state will include both locally generated wind sea, whose principal direction should be that of the local wind, and swell, i.e., long period, far travelled waves, generated up to several days earlier by distant weather patterns (which may have a quite different mean direction). In such cases, an adequate summary of the sea state will require separate heights, periods and mean directions of wind sea and (occasionally more than one) swell components. For a more precise description one can add standard deviations of period and direction for each component.

The spectral width, generally decreasing with the age of the wave systems, can be characterised by different parameters. For wave energy studies, it is preferable to use a parameter that depends

158

on the lower frequency range, where most of the energy of the sea state occurs. The standard deviation of the period σ_T (Mollison 1986) defined by

$$\sigma_T = \left(\frac{m_{-2}}{m^2_{-1}} - 1 \right)^{1/2}$$

(8)

is an appropriate parameter for that purpose because it depends on m_{-1} and m_{-2} instead of m_2 and m_4 (as other more usual spectral width parameters).

When directional spectra are not available, it is generally assumed that

$$E(f,\theta) = E(f) D(f,\theta),$$

(9)

with $D(f,\theta)$ satisfying the normalizing condition

$$\int_0^{2\pi} \int_0^{\infty} D(f,\theta) df \, d\theta = 1$$

(10)

In general, no information is available about the asymmetry of energy spreading with respect to $\theta(f)$, and a symmetric function is assumed, generally of circular type. Another simplifying assumption is to take the same spreading function for the whole frequency range $D(f,\theta) = D(\theta)$. This is a reasonable approach for sea states having only one system with mean direction θ.

One of the most widely used expressions for $D(\theta)$ is the $\cos^{2s}$ type law (Longuet-Higgins et al. 1961) defined by

$$D(\theta - \bar{\theta}) = \begin{cases} C(s) \cos^{2s}(\theta - \bar{\theta}) & \text{for } |\theta - \bar{\theta}| < \frac{\pi}{2} \\ 0 & \text{otherwise} \end{cases}$$

(11)

the normalizing $C(s)$ being given by

$$C(s) = \frac{1}{\sqrt{\pi}} \frac{\Gamma(s+1)}{\Gamma\left(s+\frac{1}{2}\right)}$$

(12)

with Γ the Gamma function (see Abramowitz & Stegun 1979). The value of s, which describes the directional spreading of the energy around mean direction θ, depends on the age of the sea state. For locally generated waves (wind sea), where a wide range of directions (and frequencies) are present, generally $s = 1$ is appropriate. With the increase of the distance to the area where that sea state (swell) was generated, the directional spreading decreases, requiring for s values of 2 to

5 or more for its description. Different procedures for selecting the value of the spreading parameter s are adopted according to the available wave information. When the energy spectra are available, a spectral width parameter such as σ_T (equation 8) can be used to select the value of s. If only mean height and period parameters are known, the choice of the directional spreading parameter s for each sea state can be based on the slope H/λ, λ being a typical wave length of that sea state.

For monochromatic waves, the relationship between λ and T is obtained from the dispersion relationship

$$\omega^2 = gk \tanh(kh).$$

(13)

where g is the acceleration due to gravity; $\omega = 2\pi f$, the circular frequency; $k = 2\pi/\lambda$, the wave number; and h is the water depth.

In deep water, $\tanh(kh) \approx 1$, thus the dispersion relationship reads

$$\omega^2 = gk$$

(14)

The slope β is then given by

$$\beta = \frac{H}{\lambda} = \frac{2\pi H}{gT^2}.$$

(15)

For irregular waves, the corresponding slope parameter can be defined as

$$\beta_e = \frac{2\pi H_s}{gT_e^2}.$$

(16)

In the case of purely wind waves (wind sea), $\beta \approx 0.004$ can be assumed. Then the significant wave height of such a sea state, denoted by $H_{s,w}$, is approximately given by

$$H_{s,w} \approx 0.06 T_e^2.$$

(17)

For a sea state having $\beta \approx 0.04$ or equivalent, $Hs = H_{s,w}$, would correspond to a $\cos^{2s}$ directional distribution, with $s = 1$. The lower H_s is with respect to $H_{s,w}$, the more it is towards swell, and the narrower the directional distribution. For $Hs = 0.1 H_{s,w}$, the $\cos^{10}$ directional distribution seems to be an appropriate choice.

Wave Energy

The wave power, or flux of energy per unit crest length, is computed by:

$$P = \rho g \int_0^{2\pi} \int_0^\infty c_g(f,h) S(f,\theta)\, df\, d\theta$$

(18)

where ρ is the water density. The group velocity c_g, i.e., the velocity at which the energy propagates, is defined by

$$c_g = \frac{\partial \omega}{\partial k}. \tag{19}$$

In deep water, c_g reduces to

$$c_g = \frac{g}{4\pi f} \tag{20}$$

thus the wave power is given by

$$P = \rho g \int_0^{2\pi} \int_0^{\infty} f^{-1} S(f,\theta)\, df\, d\theta = \frac{\rho g}{4\pi} m_{-1} \tag{21}$$

which can be expressed in terms of H_s and T_e as

$$P = \frac{\rho g^2}{64\pi} H_s^2 T_e \tag{22}$$

When H_s is expressed in meters and T_e in seconds, the power level is given by the equation

$$P_w \cong 0.5 H_s^2 T_e \ \text{kW/m}. \tag{23}$$

The energy flux incoming from the angular sector θ_k (sector centered on θ_k with width $\Delta\theta$) is computed by:

$$P_{\theta_k} = \frac{\rho g^2}{4\pi} \int_{\theta_k - \frac{\Delta\theta}{2}}^{\theta_k + \frac{\Delta\theta}{2}} \int_0^{\infty} f^{-1} S(f,\theta)\, df\, d\theta \tag{24}$$

The energy flux in a given direction θ_0, another relevant quantity wave energy utilisation, is computed by

$$P_{\theta_0} = \frac{\rho g^2}{4\pi} \int_0^{2\pi} \int_0^{\infty} f^{-1} S(f,\theta) \cos(\theta - \theta_0)\, df\, d\theta. \tag{25}$$

The most favourable direction θ_f maximizes P_{θ_0}. The directional coefficient, defined by

$$d_\theta = \frac{P_{\theta_f}}{P}, \tag{26}$$

is useful to characterise the energy spreading. Values of d_θ close to unity indicate that the energy is concentrated around θ_f, which is a favourable situation for the extraction of wave energy by systems sensitive to the direction.

References

Abramowitz, M. and Stegun, I. A. 1979 *Handbook of Mathematical Functions*; New York., NY: Dover.

Komen, G., Cavaleri, L., Donelan, M., Hasselmann.,K., Hasselmann, S. and Janssen, P.A.E.M. 1994, *Dynamics and Modelling of Ocean Waves*. Cambridge Univ. Press, 532 p.

Longuet-Higgins, M. S., Cartwright, D.E., and Smith, N.D. 1961 Observations of the direction spectrum of sea waves using the motions of a floating buoy. In *Ocean Wave Spectra*, Prentice Hall, Englewood Cliffs, N.J., 111-132.

Mollison, D. 1986 Wave climate and the wave power resource. In *Hydrodynamics of Ocean Wave-Energy Utilization* (eds. D.V. Evans & A. F. de O. Falcão), 133-156.

---

1. Contributed by Dr. Teresa Pontes, INETI-Department of Renewable Energies, Lisbon, Portugal. (Dr. Pontes is a member of the ECOR Working Group on Wave Energy Conversion.)

Appendix 3

WADIC: Directional Wave Instrumentation, Performance Evaluation
(WADIC: Wave Direction Measurement Calibration Project)

	Wave Heights	Wave Periods	Wave Directions	Wave Spread
WAVERIDER	Underestimates at high H_{rms}. Noisy at very low frequencies (0.05 Hz). Buoy may be dragged under or sensor set into oscillation in extreme waves	Overestimates T_z at high H_{rms}.	*Not applicable.*	*Not applicable.*
WAVEC	Underestimates at high H_{rms}. Stability/capsizing problems in steep waves.	Small overestimates at low periods for T_z.	OK (but engineering parameters only were available).	Small overestimates (but available only for a few days).
WAVETRACK	Underestimates at high H_{rms}. Noisy at low frequencies (14 & 25 sec resonant peaks). Mooring interference.	Overestimates T_z. Fitting high frequency tail would improve T_z.	Poor for periods > 10 sec. Generally more variable than heave/pitch/roll buoys.	Little correlation with best estimate data set (BEDS) (2).
WAVESCAN	No significant difference.	No significant difference.	Unreliable at low frequencies (<0.06 Hz) (1).	Overestimates.
MAREX	Noisy at low frequencies (<0.06 Hz) causing small H_{rms} overestimates. May have stability problems in extreme waves.	T_z overestimated for H_{rms} < 4 m, but less than 5%.	Unreliable at low frequencies (<0.06 Hz).	Overestimates but smaller differences than other buoy systems.
NORWAVE	Noisy at medium and very low frequencies causing small H_{rms} overestimates in lower sea states. Noisy time-series of heave.	T_z overestimated, but less than 5%.	Noisy time-series lead to greater variability, particularly northerly directions. Relatively high bias in wave direction. Unreliable at low frequencies (<0.08 Hz) (1).	Overestimates.
WADIBUOY	Noisy at high frequencies. Large overestimates in medium to low sea states.	T_z severely overestimated.	Relatively high variability. Poor at high frequencies (<0.3 Hz). Unreliable at low frequencies (<0.07 Hz) (1).	Overestimates.

Table 1, Performance of buoys

163

	Wave Heights	Wave Periods	Wave Direction	Wave Spread
Current Meter/Pressure Transducer Triplet (At 6 m)	Underestimates H_{rms} for low sea states. (Note: Pressure data used.)	Overestimates T_z. Fitting high frequency tail would improve T_z.	Poor at high frequencies (>0.3 Hz). Poorer than 15 m triplet at low frequencies due to marine growth. Small direction dependent offset, caused by interference from tower on which meters are mounted.	Used in BEDS (2). Overestimates swell spread. Otherwise overestimates due to noisy data.
Current Meter/Pressure Transducer Triplet (at 15 m)	Underestimates H_{rms} (5-10%) but high correlation. Systematic underestimation relative to current meter at 6 m probably due to inaccurate depth used in linear wave theory transfer function (Note: Pressure data used.)	Overestimates T_z. Fitting high frequency tail would improve T_z.	Poor at high frequencies (>0.2 Hz). (Note: Best of all systems for low frequencies.) Small direction dependent offset, caused by interference from tower on which meters are mounted.	Used in BEDS (2). Small overestimates due to noisy data.
Laser Pentagon Array	Naturally high correlated as used in BEDS (2). Ref. EMI Lasers for other comments.	Naturally high correlated as used in BEDS (3).	Unreliable at very low frequencies (<0.07 Hz), and more variable than heave/pitch/roll buoys at high frequencies.	Overestimates.
Miros Radar	Underestimates H_{rms}, particularly in low sea states. Wave spectrum accuracy strongly variable w.r.t. wave height and frequency.	Overestimates T_z. Fitting high frequency tail would improve T_z.	180° ambiguity in direction estimates.	Overestimates. Wave spread spectrum does not show characteristic dip at spectral peak as seen by all other systems.
WAVESTAFF	Underestimates H_{rms} for lower sea states (<4 m) (5%-10%)	No significant difference.	*Not applicable.*	*Not applicable.*
EMI Lasers	Naturally high correlated as used in BEDS (2). Time series somewhat noisy at times, i.e. spikes and flat tops, but not significantly affecting H_{rms}. May affect steepness parameters.	Naturally high correlated as used in BEDS (3).	*Not applicable.*	*Not applicable.*

Table 2, Performance of several wave measuring systems.
Source (both tables): T. Pontes, Energias Renovates, Portugal, based on Allender, J.T. *et al.* 1989.

[KEY (both tables): (1) directions more reliable at low frequencies for higher energy levels; (2) BEDS=best estimate data set.]

Appendix 4

WAVE PROPULSION[1]

The ability to extract energy from waves has led a number of researchers and inventors to study the possibility of wave propulsion of boats, ships and even marine mammals. One of the first attempts was documented in 1895 in the form of a patent for a self-propelled boat. The boat was equipped with a fin located at the stern. Due to the heave motion of the boat, the fin will bend upwards and downwards, propelling the boat forward. It was claimed that the boat could reach a velocity of 4 knots when moving against the waves (Anon. 1979).

Some pioneering work on the extraction of energy by a wing oscillating in waves was done by Wu (1972). This led to work first on numerical simulations (e.g. Isshiki 1982; Terao 1982); then on the propulsion of ship models in waves (e.g. Isshiki & Murakami 1983, 1984 and 1986; Kjaerland & Eggen 1980); and finally on full scale applications on boats and a small research vessel (Anon. 1983; Berg 1985). In the 1980s, Hiroshi Isshiki, Hitachi Zosen Co., Japan, carried out an extensive series of simulations and model tests that demonstrated the potential of horizontal foils as a wave propulsion device for ships (Fig. App4.1). In the early 1980s, Einar Jakobsen, Wave Control Co., Norway, demonstrated the effectiveness of a wave propulsion device, termed a foil propeller, on a 7.5 m yacht hull. This vessel reached a speed of six knots under wave power alone. Later, tests on the 20.4 m, 180 tonnes, fishing research vessel Kystfangst demonstrated that in a sea state of about 3 m wave height with horizontal bow foils (total area 3 m^2), the foils produced a propulsive force corresponding to 15-20% of the vessels total resistance at a vessel speed of 4-8 knots. Reduced pitching motion was found in head seas and reduced rolling in following seas.

Bose and Lien (1990) demonstrated numerically that an immature fin whale, Balaenoptera physalus, swimming at 2.5 ms^{-1} in a fully developed seaway corresponding to a wind speed of 20 knots, could potentially absorb between 25% and 33% of its required propulsive power in head and following seas respectively through its horizontal tail flukes. Such potential wave energy absorption has significant consequences for the energetics of cetacean migrations.

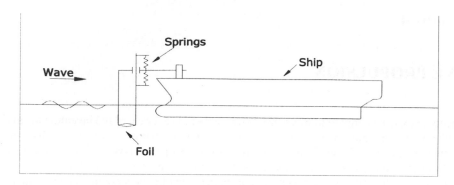

Figure App4.1 Diagrammatic representation of a wave propulsion device
Source: N. Bose, Memorial Univ. of Newfoundland, Canada. (Based on Isshiki & Murakami 1986.)
(Reprinted by permission.)

References & Bibliography

Anon. 1979 Wave energy for propelling craft - nothing new. *The Naval Architect*, Nov., 1979.

Anon. 1983 Wave power for ship propulsion. *The Motor Ship*, **64** (757), 67-69.

Berg, A. 1985 Trials with passive foil propulsion on M/S Kystfangst. *Report of Fiskeriteknologisk Forskningsinstitutt, Project No. 672.138.* Trondheim, Norway. (In Norwegian.)

Bose, N. and Lien, J. 1990 Energy absorption from ocean waves: a free ride for cetaceans. *Proc. R. Soc. Lond.*, **B 240**, 591-605.

Grue, J., Mo, A. and Palm, E. 1986 The forces on an oscillating foil moving near a free surface in a wave field. *Applied Mathematics*, **3**.

Grue, J., Mo, A. and Palm, E. 1988 Propulsion of a foil moving in water waves. *J. Fluid Mech.*, **186**, 393-417.

Jakobsen, E. 1982 Wave motors. *UK Patent*, GB 2009069B.

Jakobsen, E. 1983 Wave motor especially for propulsion of boats. *UK Patent,* GB 2045708B

Kjaerland, O. and Eggen, S. 1980 Model tests with a foil propeller – Parts I and II. *MARINTEK Report OR. 43.0 4344010.*

Korbijin, F. 1985 Analysis of a foil propeller as an auxillary propulsion system, *VERITEC Report* 85-*3326.* Hovik, Norway: VERITEC

Isshiki, H. 1982 A theory of wave devouring propulsion: 1st report – Thrust generation by linear Wells turbine. *J. Soc. Naval Architects Japan*, **151**, 54-64.

Isshika, H. 1982 A theory of wave devouring propulsion: 2nd report – Optimized foil motions for a passive type wave devouring propulsor. *J. Soc. Naval Architects Japan*, **152**, 89-100.

Isshiki, H. and Murakami, M.   1983 A theory of wave devouring propulsion: 3rd report – An experimental verification of thrust generation by a passive type hydrofoil propulsor. *J. Soc. Naval Architects Japan*, **154**, 118-128.

Isshiki, H. and Murakami, M.   1984 A theory of wave devouring propulsion: 4th report – A comparison between theory and experiment in case of a passive type hydrofoil propulsor, *J. Soc. Naval Architects Japan*, **156**, 102-114.

Isshiki, H. and Murakami, M.   1986 Wave power utilization into ship propulsion. In *Proc.5th Symposium on Offshore Mechanics and Arctic Engineering, Tokyo.*

Isshiki, H. and Naito, S.   1986 An application of wave energy – thrust generation by a hydrofoil in waves. *Energy sources technology conference and exhibition, ASME – Ocean Engr. Div., New Orleans.*

Nagahama, M., Murakami, M. & Isshiki, H.   1986 Effects of a foil attached to a ship in waves. *Hitachi Zosen Technical Review*, **37**, 38-43.

Terao, Y.   1982 A floating structure which moves toward the waves (possibility of wave devouring propulsion). *J. Kansai Soc. Naval Architects*, **184**, 51-54. (In Japanese.)

Wu, T. Y.-T.   1972 Extraction of flow energy by a wing oscillating in waves. *J. Ship Research*, **14** (1), 66-78.

---

1. Contributed by Prof. N. Bose, Memorial University of Newfoundland, Canada. (Prof. Bose is a member of the ECOR Working Group on Wave Energy Conversion.)

Appendix 5

GLOSSARY OF WAVE ENERGY TERMS[1]

A. Terms relating to the resource, the waves in the ocean.

A1. wave energy

Energy in or from waves (contrary to other energy). The total energy in a wave is the sum of potential energy, due to vertical displacement of the water surface, and kinetic energy, due to water in oscillatory motion.

A2. wave-power level, wave-energy transport

Power per unit width of the wave front. The stored energy per unit area of the sea surface multiplied by the speed of energy propagation, the so-called group velocity of the wave. SI unit for the quantity is W/m (= watt per metre) or kW/m.

(The alternative term *wave-power flux* or *wave-energy flux*, which has been used to some extent, should be avoided, because of confusion arising from the various uses of "flux" in different branches of physics and technology. A more recently proposed alternative term *wave-power level* is to be recommended.)

A3. regular wave

Wave which is periodic and has relatively long wave crests. The regular wave is closely sinusoidal and monochromatic if it is sufficiently low. Swells with long wave crests are approximately regular waves.

A4. irregular wave

Wave which is not periodic or regular. Locally generated wind sea is irregular waves. May be thought of as a set of many sinusoidal waves with different frequencies or periods and directions of propagation. See *wave-energy spectrum* and *directional spectrum*.

A5. swell

Wave that has propagated out from the region of wind generation.

A6. wavelength

The distance between two consecutive wave crests measured in the direction of wave propagation.

A7. wavefront

An envisaged plane which is perpendicular to the direction of wave propagation, and which moves with the propagation speed (phase velocity) of the wave.

A8. wave height

The vertical distance H between a wave crest and the previous wave trough

A9. significant wave height

A characteristic wave height to describe a wave state. Traditionally it is defined as the average of the largest one third of individual waves in a particular series of measured waves. It is usually approximately equal to the modern definition: 4 times the square root of the zero order moment of the wave-energy spectrum.

A10. hundred-year wave

The wave-height that on average is met or superseded once in a hundred years. (Note that the probability is 63% that this wave height will be attained at least once in a hundred years.)

A11. zero-crossing period

The time between two consecutive zero-down-crossings (events where the water level on the measurement location passes the still-water level in the downwards direction).

A12. wave-energy spectrum

A mathematical or graphical description of how a wave state of irregular waves is distributed among the various frequencies.

A13. directional wave spectrum

A two-dimensional spectrum that shows how the wave energy is distributed between various directions of incidence, in addition to how it is distributed among various frequencies. (See wave-energy spectrum.)

A14. n-th order moment of wave spectrum

A power (n) of the frequency multiplied by the wave-energy spectrum and integrated over all frequencies.

B. Terms related to structures in the sea.

B1. heave

Linear oscillatory motion (translation) of an immersed body or structure in the vertical direction.

B2. surge

Linear oscillatory horizontal motion of an immersed body in the direction of longest extension. If the body, such as an axisymmetric body, has no particular longest horizontal direction, the direction for surge motion may be specified as the direction of wave incidence.

B3. sway

Horizontal linear oscillatory motion perpendicularly to the surge motion.

B4. roll

Rotary oscillatory motion around a horizontal axis in the direction of longest extension of the immersed body, or alternatively, in the direction of wave incidence.

B5. pitch

Rotary oscillatory motion around a horizontal axis in the direction perpendicular to the axis of rotation for roll motion.

B6. yaw

Rotary oscillatory motion around a vertical axis.

B7. added mass

When a solid body oscillates harmonically (sinusoidally) in water, there will, because of the corresponding oscillatory motion of the water surrounding the body, be set up a dynamical reaction force, which has a component in phase with the acceleration of the body. The hydrodynamically *added mass* is defined as the ratio between this force component and the acceleration of the body.

B8. damping coefficient

Radiation resistance, or possibly the sum of radiation resistance and other mechanical resistance due to friction and viscous effects.

B9. radiation resistance, added damping coefficient

A hydrodynamical damping coefficient or mechanical resistance which is a measure of an oscillating body's ability to generate waves. The double ratio between the power which radiates outwards and the square of the velocity amplitude of the body when it oscillates harmonically (sinusoidally).

171

B10. excitation force, exciting force

The force which an incident wave exerts on a body, when it is not moving.

B11. wave loading, wave load

The forces which waves exert on floating, submerged or bottom-standing structures.

C. Terms related to utilisation of ocean-wave energy.

C1. wave power

Mechanical power (energy per unit time) from waves.

C2. absorbed [wave] power

The power which an oscillating system removes from the waves.

C3. radiated power

The power which is removed from the oscillating system and carried away with the radiated wave which is generated by the oscillating system.

C4. useful power, captured power

The net (useful) power which is delivered by a wave-energy converter. The difference between absorbed wave power and power that is lost due to dissipative effects, such as friction and viscosity, etc.

C5. absorption width

A measure for a wave-power device's ability to absorb power from a wave. The ratio between absorbed power and the wave-power level.

C6. capture width

A measure for a wave-power device's ability to capture power from a wave. The ratio between captured power and the wave-power level.

C7. phase control

Method to obtain optimum oscillatory motion in order to capture a maximum of wave energy. For a simple (single-mode) oscillating system the object is to obtain an oscillatory velocity that is in phase with the excitation force due to the incident wave.

C8. wave-power device, wave-energy converter (WEC)

A technical device or system designed to convert wave energy to electrical energy or another kind of useful energy.

C9. wave power plant

Power plant run by wave energy.

C10. wave-powered generator

Electrical generator run by wave energy.

C11. power buoy

Wave-power device where energy is captured by means of a buoy which performs vertical oscillation with respect to a fixed point, e.g. an anchor on the sea bed, or with respect to a submerged body which has an oscillatory motion for which the amplitude and/or phase are/is essentially different from those/that of the buoy's motion.

C12. point absorber

Wave-power device for which the horizontal extension is very small compared to predominant wavelengths, and for which the ability to absorb (and/or radiate) wave energy is essentially independent of the direction of wave incidence.

C13. line absorber

Contrary to point absorber: Wave-power device for which the longest horizontal extension is is at least as long as the predominant wavelengths.

C14. terminator

Line absorber which is alined perpendicularly to the predominant direction of wave incidence.

C15. attenuator

Line absorber which is alined along with the predominant direction of wave incidence.

C16. spine, backbone

Relatively stiff structure for which the longest horizontal extension is at least as long as predominant wavelengths, and which serves as a common reference against which the differently oscillating members of a line absorber can react.

C17. duck, Salter duck

A wave-power device of the terminator type: on a long cylindrical spine an array of duck-like bodies are mounted; where the relative motion between each "nodding-duck" body and the common spine is utilised for running pumps, which (in the first step) convert wave energy into hydraulic energy.

C18. raft, Cockerell raft

A wave-power device composed of at least two rafts or pontoons, which are hinged together. Hydraulic energy is produced through the relative oscillatory (rotary) motion between the rafts.

C19. oscillating water column (OWC)

A wave-power device with a chamber which has an opening in the water (in the sea), such that water, in interaction with the waves, oscillates in and out through the opening and makes an interface between water and air oscillate up and down within the chamber. Energy conversion may take place e.g. thereby that this interface acts as a piston of an air pump which runs an air turbine.

C20. Masuda buoy

Floating buoy containing a battery which is charged by an electic generator run by a wave-driven air turbine, through which air is driven by an OWC contained within a vertical tube, which is part of the buoy structure, and which is open in its lower end.

C21. clam

A wave-power device of the terminator type. It is composed of flexible bags or membranes on a common spine structure. Entrapped air is pumped back and forth through air turbines when the membranes oscillate under wave action outside the membrane. Instead of arranging the bags on a long straight spine, a clam device has been proposed where the bags are arranged on a circular spine.

C22. Bristol cylinder

A wave-power device of the terminator type. Cylinders, which are arranged after eachother along a line, and which by means of moorings to seabed mounted hydraulic pumps are held in a completely submerged position closely below the sea surface, and arranged to oscillate (under wave action) to move in circles around an eccentric axis parallel to the cylinder axis. Wave energy is converted to useful hydraulic energy by means of the pumps.

C23. wave-energy rectifier, Russel rectifier

Wave-energy device which is located on the sea bottom, preferably near land, and which has an upper and a lower water reservoir. Check valves allow for water to flow into/out from the upper/lower reservoir at the occurrence of a wave crest/trough. A low-head water turbine utilises the head between the two reservoirs.

C24. tapered channel, Tapchan

Horizontal tapered channel placed on the shore (or on the sea bed nearshore) such that the wide end faces the waves incident towards the shore. As waves enter and propagate along the narrowing channel, the wave height increases gradually, and as a result water is spilt from the wave crests over the upper channel edge into a water reservoir, where a water head is achieved in order to run a water turbine.

C25. ocean-wave lens

Structure(s) immersed in the sea in order to refract the waves such that they converge to a focal area analogous to the focal point of an optical lens.

C26. ocean-wave focusing

Concentration of ocean waves into certain areas or focal regions. The cause of this is that waves change their direction of propagation because of the bathymetry (sea-bottom topography) and/or because of artificial ocean-wave lenses.

C27. flexible bag

Flexible bags or membranes which are mounted on a wave-power structure in order to separate oscillating water on the outside from air on the inside, where the air us used as a driving medium for a turbine.

C28. hose pump, tube pump, Petro pump

Relatively long, specially reinforced, rubber hose which changes volume when elongated. A pump which is operated with linear oscillatory motion, although it is without a piston.

C29. Wells turbine

Air turbine which is self-rectifying, that is, its sense of rotation is the same for both of the two oppsite air-flow directions. The turbine resembles a propeller, where the blades, which have an air-wing-like profile, have no pitch. They are symmetrical with respect to the air flow's inlet and outlet sides.

---

1. Based on "Terminologi for havbølgjeenergi: Ei lita ordbok ved [Terminology for ocean-wave energy: A small glossary by] Johannes Falnes" (Trondheim, 1984). (Note: Prof. Falnes is a member of the ECOR Working Group on Wave Energy Conversion.)

Appendix 6

FURTHER READING ON THE SUBJECT

Presented below is a list of selected items for further reading on the subject of wave energy conversion. The selection, which was made by members of the ECOR Working Group, provides items at different levels of complexity. Most of these items are also included in the "References Cited" chapter, to which the reader is referred for additional papers on specific aspects of the subject.

Easy reading aimed at the general public:

- - Ross, David. 1979 *Energy from the Waves*, 1st. Edition. Oxford: Pergamon Press.
- - Ross, David. 1981 *Energy from the Waves*, 2nd. Edition. Oxford: Pergamon Press.
- - Ross, David. 1995 *Power from the Waves*. Oxford University Press. (ISBN 0-19-856511-9).

Books of a more scientific/ technical nature:

- - Claeson, Lennart et al. 1987 *Energi från havets vågor* (Energy from the ocean's waves). (In Swedish). Efn-rapport nr. 21. Stockholm: Energiforskningsnämnden.
- - Count, B. (editor) 1980 *Power from SeaWaves*. Academic Press.
- - Falnes, J. 2002 *Ocean Waves and Oscillating Systems: Linear Interactions Including Wave-energy Extraction*. UK: Cambridge University Press. (ISBN 0-521782112).
- - McCormick, Michael E. 1981 *Ocean Wave Energy Conversion*. New York: John Wiley & Sons.
- - Shaw, Ronald. 1982 *Wave Energy: A Design Challenge*. Chichester, UK: Ellis Horwood Ltd.

CD-ROMs:

- - Falnes, J. & Hals, J. 2002 Wave Energy and its Utilisation: A contribution to the EU Leonardo pilot project 1860. In: *Alter Eco, Sustainable Technology and Regenerative Energy*. Brussels: European Union (In preparation.)

Articles in encyclopedia volumes:

- - Hagerman, G. 1995 Wave Power. In: *Encyclopedia of Energy Technology and the Environment* (eds. A. Bisio & S.G. Boots), 2859-2907. New York: John Wiley & Sons Inc.

Review books on renewable energy containing chapters on wave energy:

- - G. Boyle (ed.). 1996 *Renewable Energy. Power for a Sustainable Future.* Oxford University Press.(ISBN 0-19-856451-1).
- - Charlier, R.H. & Justus, J.R. 1993 Waves. In *Environmental, Economic & Technological Aspects of Alternative Power Sources*, Chapter 5. Elsevier Oceanography Series, No. 56.
- - Jackson, T. (ed.). 1993 *Renewable Energy: Prospects for implementation.* Stockholm Environment Institute. (ISBN 91 8811672-1).
- - Seymour, R.J. (ed.). 1992 *Ocean Energy Recovery. The State of the Art.* New York: American Society of Civil Engineers, New York. (ISBN 0-87262-894-9).
- - Taylor, R.H. 1983 *Alternative Energy Sources for the Centralised Generation of Electricity.* Bristol, UK: Adam Hilger, Ltd.
- - Twidel, J.W. & Weir, A.D. 1986 *Renewable Energy Resources.* London: E.& F.N. Spon Ltd.
- - Wick, G.L. & Schmitt, W.R. 1981 *Harvesting Ocean Energy.* Paris: UNESCO.

Proceedings of conferences on wave energy:

- - *Proceedings of International Symposium on Wave and Tidal Energy, Proceedings, 27-29 September, 1978, Canterbury, UK.* (ISBN 0-906085-00-4).
- - *Proceedings of First Symposium on Wave Energy Utilization, 30 October-1 November, 1979, Goteborg, Sweden.* Göteborg, Sweden: CTH.
- - *Proceedings of Second International Symposium on Wave and Tidal Energy, 23-25 September, 1981, Cambridge, UK.* (ISBN 0-906085-43-8).
- - *Proceedings of Second International Symposium on Wave Energy Utilization, 22-24 June, 1982, Tapir, Trondheim, Norway.* (ISBN 82-519-0478-1).
- - *Proceedings of IUTAM Symposium on Hydrodynamics of Ocean Wave Energy Utilization, 1985, Lisbon, Portugal* (eds. D.V. Evans and A.F.O. Falcão). Springer Verlag. (ISBN 3-540-16115-5 or 0-387-16115-5).
- - *Proceedings of Third Symposium on Ocean Wave Energy Utilization, January 22-23, 1991, Tokyo, Japan* (eds. T. Miyazaki and H. Hotta). Japan Marine Science and Technology Center.
- - *Proceedings of International Symposium on Ocean Energy Development, 26-27 August, 1993, Muroran, Hokkaido, Japan.* (ed. H. Kondo). (ISBN 4-906457-01-0).
- - *Proceedings of 1993 European Wave Energy Symposium, Edinburgh, Scotland, 21-24 July, 1993* (eds. G. Elliot and G. Caratti). Brussels: European Commission, ECSC-EEC-EAEC, 1994. (ISBN 0-903640-84-8).
- - *Proceedings of Second European Wave Power Conference, Lisbon, Portugal, 8-10 November, 1995* (eds. G. Elliot & K. Diamantaras). Brussels: European Commission, ECSC-EC-EAEC, 1996 (ISBN 92-827-7492-9).

178

- - *Proceedings of Third European Wave Energy Conference, Patras, Greece, 30 September-2 October, 1998* (ed. W. Dursthoff). University of Hannover, 2000.
- - *Proceedings of Fourth European Wave Power Conference, Aalborg, Denmark, 4-6 December, 2000* (eds. I. Østergaard & S. Iversen), 243-250. Tåstrup, Denmark: Danish Technological Institute, 2001 (ISBN 87-90074-09-2).

Appendix 7

WEB SITES OF INTEREST

Web sites that describe the programs of specific countries, agencies and companies are for the most part included in the text of the book and listed in the "References Cited" chapter. The following sites are of a more general nature:

(1) *Title*   Alternative Energy Sources- Wave Energy
                    (Johannes-Kepler-Gymnasium Reutlingen, Germany)
     Web Site: http://www.schwaben.de/home/kepi/waves3.htm
     Description: Simple descriptions, with diagrams, of: OWC; surge device; heaving &
                    pitching device; pitching device

(2) *Title:*   EREN - Wave Energy Converters - Energy Efficiency and Renewable
                    Energy
     Web Site: http://www.doe.gov/html/eren/1608.html
     Description: Network. Subject Category: Tidal and Wave Power / Wave Energy
                    Converters. WAIS Search of Current Level.

(3) *Title:*   European Wave Energy Research Network- JOULE
     Web Site: http://erg.ucd.ie/thermie.html
     Description: Lists JOULE Information. Direct link to the JOULE (-THERMIE) pages.

(4) *Title:*   European Wave Energy Atlas (WER-Atlas)
     Web Site: http://www.ineti.pt/ite/weratlas
     Description: WER-Atlas page, INETI. Participants in the WER-Atlas project.

(5) *Title:*   European Wave Energy Research Network
     Web Site: http://www.ucc.ie/ucc/research/hmrc/ewern.htm
     Description: News items.

(6)   *Title:*   Tidal and Wave Energy Newsletter Articles - Tidal and Wave Energy
                        Newsletter
       Web Site: http://www.caddet.co.uk/nlttdwav.htm
       Description: Articles, Issue Number and Title. e.g. 2/94 The UK Tidal Energy
                        Programme; 2/95 Wave Power for Propulsion - Norway.

(7)   *Title:*   U.S. Wave Energy Data
       Web Site: http://194.178.172.97/class/ixb14.htm
       Description: Ocean Wave Energy. Source Country for Data: United States
                        Date: August, 1994. Technology Data Type: Best Available Practice.

(8)   *Title:*   Wave Energy Activities at OCEANOR (i.e. Oceanographic Company of
                        Norway)
       Web Site: http://oblea.oceanor.no/wave_energy/
       Description: Wave energy pre-feasibility and resource studies carried out around the
                        world over the past 15 years.

Appendix 8

ACRONYMS AND ABBREVIATIONS

AG	Asynchronous Generator
ART	Applied Research and Technology (UK)
BBDB	Backward Bent Duct Buoy
BODC	British Oceanographic Data Centre
CCGT	Combined Cycle Gas Turbine
CONWEC	Controlled Wave Energy Converter
DOE	Department of Energy (USA)
DTI	Department of Trade and Industry (UK)
EC	European Community
	European Commission
ECMWF	European Centre for Medium-Range Weather Forecasts
ECN	Ecole Centrale de Nantes
ECOR	Engineering Committee on Oceanic Resources
EEC	European Economic Community
EPRI	Electric Power Research Institute (USA)
ESA	European Space Agency
FWPV	Floating Wave Power Vessel
IEA	International Energy Agency
IFREMER	Institut Français de Recherche pour l'Exploitation de la Mer (France)
INETI	National Institute of Engineering and Industrial Technology (Portugal)
ISBN	International Standard Book Number
ISOPE	International Society of Offshore and Petroleum Engineers
IUTAM	International Union of Theoretical and Applied Mechanics
JPD	Joint Probability Distribution
IPS	Interproject Service (Sweden)
JAMSTEC	Japan Marine Science and Technology Center
JONSWAP	Joint North Sea Wave Project
KRISO	Korean Research Institute of Ships and Ocean Engineering
LIMPET	Land Installed Marine Pneumatic Electrical Transformer
MWP	McCabe Wave Pump
NEL	National Engineering Laboratory (UK)
NTNU	Norwegian University of Science & Technology
OPT	Ocean Power Technologies (USA)
OSPREY	Ocean Swell Powered Renewable Energy
OTEC	Ocean Thermal Energy Conversion

OWC	Oscillating Water Column
OWEC	Offshore Wave Energy Converter
PACON	Pacific Congress on Marine Science and Technology
QUB	Queen's University Belfast (UK)
RO	Reverse Osmosis
SAR	Synthetic Aperture Radar
SEA	Sea Energy Associates (UK)
SG	Synchronous Generator
SOPAC	South Pacific Applied Geoscience Commission
TAPCHAN	Tapered Channel
WACSIS	Wave Crest Sensor Intercomparison Study
WADIC	Wave Direction Measurement Calibration Project
WAM	Wave Modeling Group
WEC	Wave Energy Converter
	World Energy Conference
WEM	Wave Energy Module
WERATLAS	Wave Energy Resource Atlas (Europe)
WG	Working Group

SUBJECT INDEX

Archimedes Wave Swing (Netherlands), 107

Backward Bent Duct Buoy (Japan), 79, 84 87, 90, 92
Bristol Cylinder, 40-41

Contouring Float System (Japan), 87
Contouring Raft System (UK & US), 73, 115
costs
        capital costs, 48-49
        comparison, other renewables, 52-54
        comparison, conventional generation
                costs, 53-54
        comparison, electricity prices, 53-54
        costs for specific devices, 50
        external costs, 55
        fixed charges, 48
        generating costs, 49-50
        levelized cost of energy, 47
        methodology, 47
        operating costs, 48-49
        parametric costing, 49
        variable costs/expenses, 48
cyclones, extra-tropical, 15

Danish Heaving Buoy, 104
Dawanshan Island OWC (China), 89
DELBUOY (US), 70-71
desalination, 3, 55, 65
design selection, 6

economics (*see* costs)
Edinburgh Duck (UK), 118
electric power conversion
        asynchronous generators, 44-45
        electronic power conversion, 45
        linear generators, 44
        multi-pole generators, 44
        synchronous generators, 44-45
energy storage, short-term

flywheels, 43
        gas accumulators, 43
        water reservoirs, 43
environmental benefits, 57-59
environmental impact assessment, 63-64
environmental impacts
        amenity loss, 60
        coastal deposition/erosion, 60-61
        ecosystems, 61
        fishing, 62
        marine pollution, 63
        shipping, 62
        recreation/tourism, 62-63
        summary, 5, 58

Floating Wave Power Vessel (Sweden, UK), 101, 120
flywheels, 43
fossil-fuel power plants, environmental penalties, 55-56

Hose-Pump (Sweden), 72, 101-102
hydraulic systems, 42-43

industrial benefits, 67
IPS Buoy (Sweden, US), 74, 102-103, 118

Kaimei Floating Platform (Japan), 69, 79, 83
Kaiyo Jack-up Rig (Japan), 87
Kaplan turbine, 42
Kujukuri OWC (Japan), 81

Lancaster Flexible Bag (UK), 118-120
latching control, 40
legal issues, 67
LIMPET OWC (UK), 110
levelized revenue requirement methodology, 47

Masuda Buoy (Japan), 3, 27, 79, 82, 89
McCabe Wave Pump (Ireland), 120-122

Subject Index

atlases, 15, 20, 25
bottom friction, 8
breaking, 8
climatology, 15, 20-21
data, 18, 24-25
diffraction, 9, 22
dispersion, 13
extreme, 18
height, 13, 20, 23
hydrodynamics, 157-160
mathematical description, 157-162
period, 13, 20
power levels, 7, 11, 13, 15, 18-19, 69,
79, 100, 125
refraction, 8, 10, 22
variability, 13
visual observations, 19-20
Wave Dragon (Denmark), 105
wave energy,
electricity supply, 2
mathematical approach, 7, 160-161
period, 11
resource assessment, 21, 25-26
resources, 15
resources, global, 2, 7
storage, short-term, 43
wave energy applications
aquaculture, 66
desalination, 65
electricity supply, 65
in isolated coastal communities, 65
in remote island communities, 65
sea-water renewal, 66-67
wave energy conversion
attenuator, 28
wave energy absorption, 30-31
energy converter, 30-31
optimization, 31-32
primary conversion, 27-28
process, definition, 27
processes, classification, 28-30
secondary conversion, 27-28
terminator, 28
wave energy converters (*see also specific
devices*)
listing of devices, 32-35

location zones, 32
operational status, definitions, 32
Wave Energy Module (US), 71-72
wave measurements
in situ, 21
remote sensing, 22
wave measurement systems
aerial photography, 22
acoustic probes, 21-22
buoys, 21-22, 163
current meters, 21
directional, 22
orbital velocity sensors, 21
performance, 163-164
shore-based radar, 23
synthetic aperture radar (SAR), 23
satellite altimeter, 23-24
wave staff, 21-22
Wave Mill (Denmark), 106
Wave Plane (Denmark), 105
Wave Plunger (Denmark), 106
wave power schemes, worldwide, 2
wave-powered desalination, economics, 55
Wave Pump (Denmark), 106
wave-rider buoy, 19
wave rotor, 37-38
wave statistics
long-term time series, 15
mathematical approach, 157-162
Wave Turbine (Denmark), 106
Wells Turbine, 38-39, 60, 83, 92
Wind-Wave Power Device (S. Korea), 51, 53,
88
winds
extra-tropical storm, 17
measurement, 23

187